MW00608425

Conditionals

Conditionals

Nicholas Rescher

A Bradford Book
The MIT Press
Cambridge, Massachusetts
London, England

© 2007 Massachusetts Institute of Technology

All rights reserved. No part of this book may be reproduced in any form by any electronic or mechanical means (including photocopying, recording, or information storage and retrieval) without permission in writing from the publisher.

MIT Press books may be purchased at special quantity discounts for business or sales promotional use. For information, please e-mail special_sales@mitpress.mit.edu or write to Special Sales Department, The MIT Press, 55 Hayward Street, Cambridge, MA 02142.

This book was set in Times New Roman and Syntax on 3B2 by Asco Typesetters, Hong Kong and was printed and bound in the United States.

Library of Congress Cataloging-in-Publication Data

Rescher, Nicholas.
Conditionals / by Nicholas Rescher.
  p. cm.
"A Bradford Book."
Includes bibliographical references (p.) and index.
ISBN-13: 978-0-262-18259-1 (hardcover : alk. paper)
1. Conditionals (Logic). 2. Knowledge, Theory of. I. Title.
BC199.C56R47 2007
160—dc22   2006027951

10 9 8 7 6 5 4 3 2 1

For Gereon Wolters

Der Mann der das Wenn und das Aber erdacht, Hätt' sicher das Häckchen zu Gold schon gemacht.

Who if and but devised of old, would make mere straw turn into gold.
—German proverb

Contents

Preface

The present book seeks to clarify the idea of what a conditional says by way of elucidating the information that is normally transmitted by its utterance. It was projected in March of 2001, drafted over the rest of that year, and polished during the next. Though a part of it—specifically that dealing with counterfactuals—goes back to ideas developed over many years, its treatment of conditionals in general represents a newly devised uniform approach to conditionals at large through their across-the-board reduction to logico-conceptual entailment ($\vdash$). As a result, what we have here is a unified treatment of conditionals based on epistemological principles rather than upon the semantical principles that have been in vogue over recent decades. As the book's deliberations will make manifest, such an approach makes it easier to understand how conditionals actually function in our thought and discourse.

As regards the book's treatment of counterfactuals, it should be stressed that the present treatment is in essence a revival of an epistemic approach adumbrated by F. P. Ramsey in the 1920s and developed by myself in the early 1960s. It was subsequently overshadowed by the popularity of a possible worlds approach incorporated by a 1968 paper by Robert Stalnaker. However, the liabilities of the semantical possible worlds strategy have become increasingly clear over the intervening years and the time now seems ripe to reappraise the promise and utility of this older epistemological and pragmatic mode of analysis.

I am grateful to Alex Pruss and Samuel Wheeler III for many constructive comments on an early version of the book, and I also want to acknowledge the excellent work of Estelle Burris in preparing the typescript.

Pittsburgh, Pennsylvania
December 2002

Conditionals

1 Fundamentals

1 Conditionals

"Iffy" thinking is one of the characteristic resources of the sorts of creatures we humans have become. For in intellectual regards, the *Homo sapien* is an amphibian who lives and functions in two very different realms—the domain of existing reality that we can investigate in observational inquiry, and the domain of suppositional projection that we can explore in creative imagination. And this second ability becomes crucially important for the first as well, when once one presses beyond the level of a mere *description* of the real to concern oneself also with its rational *explanation*, which involves providing an account of why things are as they are rather than otherwise. In the history of Western thought this transition was apparently first made by the Greek nature-philosophers of pre-Socratic times, who, as best we can now tell, invented "thought experimentation" as a cognitive procedure and—as will be seen—practiced it with great dedication and versatility.

A thought experiment is an attempt to draw instruction from a process of hypothetical reasoning that proceeds by eliciting the consequences of a hypothesis which, for aught that one actually knows to the contrary, may well be false. It consists in reasoning from a supposition that is not (or not as yet) accepted as true—and perhaps is even known to be false—but is assumed provisionally in the interests of making a point or establishing a conclusion.[1]

In natural science, thought experiments are commonplace. One classic instance is Albert Einstein's pondering the question of what the world would look like if one were to travel along a ray of light. Think too of the physicists' assumption of a frictionlessly rolling body or the economists' assumption of a perfectly efficient market and similar devices for establishing the laws of descent or the principles of exchange. Indeed, thought experiments are far more common in science than one may think. More than a century ago, Ernst Mach made the sound point that any sensibly designed *real* experiment should be preceded by a *thought* experiment that anticipates at any rate the possibility of its outcome.[2] The conclusion of *such* a thought experiment will clearly be hypothetical: "If the experiment turns out X-wise, we shall be in a position to conclude...." There is good reason to see such thought experimentation as an indispensably useful accompaniment to actual experimentation.[3]

To us moderns, brought up on imaginative children's nursery rhymes ("If wishes were horses, then beggars would ride") and accustomed to obvious adult fictions, this sort of belief-suspensive thinking seems altogether natural. It takes a logician to appreciate how complex and sophisticated thought experimentation truly is. For what it involves is *not* simply the drawing of an appropriate conclusion from a given true fact, but also the higher-level consideration that a particular thesis (be it fact or mere supposition) carries a certain conclusion in its wake. Thought experimentation thus takes us from mere inference to hypothetical thinking at large.

Hypotheses open up a gateway to conditional reasoning, and conditionality covers a vast range. There are not just conditional assertions but also conditional questions ("If winter comes, can spring be far behind?"), conditional commands ("If he comes late, give him hell!"), conditional directions ("If you just keep going straight, you'll find the town hall"), conditional promises ("If you do the work, you will be well paid for it"), conditional advice ("If ever in doubt, ask for directions"), and more. However, the conditionals that will primarily concern us here are assertoric: they are statement-connective statements that are explicitly or implicitly of "if-then" form, for ordinarily the consequent of the conditional spells out what follows (in one of the many senses of this term) from the acceptance or supposition of the antecedent. "If he comes, then I shall leave" is a typical example.

As regards nomenclature, with an assertorically consequential conditional of the format "If p, then q" (symbolically $p \Rightarrow q$) the first, "iffy" part (p) is called the *antecedent* (protasis), and the second, thenny part (q) is call the *consequent* (apodosis). To be sure, it must be granted that not every if-then statement is genuinely conditional. For example, some such statements convey a mere expression of politeness:

If I may offer a suggestion, then do X.

What we have here is simply a more tentative—and thereby more polite—way of making the suggestion at issue. Analogously,

If you need money, there is plenty in your account

is not really a proper conditional, although it could be reformulated as something like "If you need to obtain money, then you can succeed in doing so by taking it from your account (where there is plenty of it)."[4] Such statements are cast in conditional form but do not, as they stand, involve the consequential idea that one thing follows from or upon another. Assertorically consequential conditionals, by contrast, spell out a result of projecting a certain assumption or supposition. In taking the form "If p, then q," they set out the claim that (though, of course not *how*) q is a consequence that somehow follows in the wake of the hypothesis p. The present book will be concerned only with such consequential statements.

The range of assertorically conditional discourse is surprisingly extensive. Of course, there are the obvious "if-then" propositions. But conditionals can wear a disguise in this regard. "Should he come, I'm out of here" comes sufficiently close to count as a variantly formulated conditional. And "This button operates that light" is effectively equivalent with "If you push this button, that light will switch its on/off state." "Whenever it rains, it pours" comes to: "If it rains, it also pours" and "Wherever he goes, they follow him" comes to: "If he goes someplace, they follow him." "Membership in the club requires a fee" comes to "If someone is to join the club, then this individual must pay a membership fee." Again, "Doing X successfully requires a great effort" in effect comes to "If one is to succeed at doing X, one will have to exert a great effort." Even merely descriptive statements can carry an essentially conditional message. "Dogs are mammals" to all intents and purposes comes to "If a creature is a dog, then it is a mammal." Thus the aspect of conditionalization is often concealed. Consider: "No American adult, even the least well uninformed, can fail to realize that George Washington was the first president," which amounts to the doubly conditionalized statement:

If someone is an American, then even if he is among the least informed he will (nevertheless) realize that George Washington was the first president.

The language of provisos also provides examples of obliquely formulated conditionals, with "provided" as *if* or possibly *iff*. Again, "Barring unforeseen developments he will be joining us" comes to "If unforeseen developments do not arise, he will be joining us." And "Should they invite him, he'll of course accept" is a further example of conditionals that sidestep the standard if-then formulation. Then too, in American usage conditionals can assume an if-then-lacking format of the structure:

Present tense + and + future tense

For example:

You do that, and I'll never speak to you again.

He comes in wearing those clothes, and they'll throw him out.

More generally, the construction "when + future tense" is sometimes used for "if":

When you look into it, you'll find that this can also be done in other ways.

Here there is no predictive claim that you actually are going to look into it in the future. However, there are also contexts where "when" is strictly time-indicative and is something very different from "if" because there is no intention to assert anything iffy or uncertain. A passenger would be dismayed at an iffy reformulation of "When we land, the pilot will turn off the 'Fasten Seat Belt' sign." And a patient would not be happy to have his doctor substitute "if" for "when" in the otherwise reassuring "When you recover from this ailment you'll be good as new."

The long and short of it is that assertoric conditionality is not so much a grammatically taxonomic category as a functional category in the domain of information management. "Had he but known they would abandon him, he would not have trusted them" is every bit as much of a conditional statement as "If he had known they would abandon him, then he would not have trusted them."

What has been called "Geach's thesis" is the contention that only a completed proposition can function as the antecedent of a meaningful conditional.[5] This contention may seem problematic in the face of such conditionals as

If not now, then never.

If awake, he will not have failed to notice.

But this reservation could be countered that such statements are merely elliptical, and that what is really at issue is a rather different conditional:

If we do not do the deed now, then we shall never do so.

If he was awake at the time, then he will not have failed to notice.

However, a graver problem for Geach's thesis arises with the subjectivity exhibited by such subjunctive conditionals as:

If Jones were here, this mess could not have arisen.

Here the antecedent scarcely qualifies as a completed statement, for while "Jones was here" indeed qualifies, "Jones were here" does not, seeing that it fails to state. More damaging yet for Geach's thesis are conditionals involving propositional functions rather than propositions. Take

If an integer is greater than 5, then it is greater than 4,

or, symbolically, $(\forall x)(x > 5 \Rightarrow x > 4)$. There is simply no way to recast this conditional into the format $p \Rightarrow q$, with p as a "completed proposition."

2 Supposition, Hypothesis, Assumption

Informatively consequential conditionals are answers to "what-if" questions: "What if it rains tomorrow?," "What if dinosaurs had never evolved?," "What if a number equals twice its square root?" The object of such conditional assertions is to let our thought and discourse move above and beyond the range of categorical actuality. They enable us to say something about what must, will, or can be in circumstances that we merely assume rather than believe. Without conditionals our thought would be restricted to reality—constrained to the decidedly limited, factual range of what is actually so. Speculation, planning, and conjecture would be aborted.

We can, to be sure, inquire into the reality of things, but we also have at our disposal an imagination that affords the resources of assumption, supposition, hypothesis, and conjecture. And the main reason for preoccupation with the theory of conditionals is their role in broadening the range of reason. The crucial fact of it is that we can reason equally well both from fact and from supposition. We can say:

Since he is here, we can give the book to him.

But we can also—and no less appropriately—say:

If he were not here, we could send the book to him.

Thus the antecedent of a conditional need not be known; it can merely be supposed. Conditionalization is intimately intertwined with hypothetical reasoning. Since, by definition, one does not propose to assert the premises of a hypothetical inference as true, its status within the framework of our "knowledge" will be that of a *supposition*. And this term will here be used in a very broad sense, to apply to any informative statement put forward within what may be called a *supposition-context*, illustrated by such qualifying phrases as:

Let us suppose that...

Let us assume that...

Let it, for the sake of argument, be agreed that...

Let it, for the sake of discussion, be accepted that...

Let the hypothesis be made that...

Let...be so

The close relationship between conditional assumption and hypothetical inference lies at the heart of the present deliberations. For whenever we obtain a conclusion by drawing an inference from a supposition—no matter how elaborate the course of reasoning may be—then we can summarize the overall result by if-then conditionalizations. As the ensuing discussion will show, assertoric conditionals are always, in effect, reports on what in some suitable sense follows by inference from the substance of such an assumption or supposition.

Speculatively fact-dismissive conditionals are not limited to saying what *would* happen when or if but can just as readily deal with merely speculative possibility, with what *could or might* happen: "If you have a license, then you can drive." Then too they can deal with normative or evaluative issues, with "what ought to happen if," or "what it would be nice to happen if." The conditional

If (only!) you had spoken up, you might have prevented it

is every bit as authentic a conditional as

If you had spoken up, this would have prevented it.

Such conditional possibilities and necessities are often both informative and useful. For example, the conditionals

If you take an umbrella, then you can keep yourself dry with it

If they took the 6 AM train, then they must be home by now

convey potentially helpful information.

If-then thinking of the sort involved in imaginative wondering, planning, promising, and various other activities is a pervasive and characteristic mode of human endeavor. We live the whole of our physical existence in the realm of what is. But much of our mental existence is lived in the realm of what might be.

3 Enthymematic Bases

"If you open that box, you'll find a gold ring in it" actually says little more than "There is a gold ring in that box." In general, when a conditional that claims "If p then q" is appropriate, there lies in the background some categorical (unconditional) facts in virtue of which this conditional obtains—facts that must obtain for the conditional to hold. This body of fact is not, however, something that the conditional explicitly asserts. It is the tacit, *enthymematic basis* on which the appropriateness of the conditional rests, although the conditional itself usually does no more than hint at what this basis of underlying fact actually is. When one asserts a conditional this enthymematic basis must belong to the manifold of one's beliefs: it must be accepted or at any rate supposed to be true. Were it to be regarded otherwise—as a merely speculative possibility, say—then that conditional is not in order. And likewise in the case of the counterfactual conditional, in particular, however much the antecedent may be regarded as fanciful and false, the enthymematic basis of the conditional must be deemed true. Consider, for example:

If Queen Elizabeth I had married, she would have had a husband.

The marriage of Elizabeth I is clearly something fanciful and unreal. But the background presence of something on the order of the consideration that a married woman takes on a husband in the process—at least for a time, however brief—has to be seen as a fixed fact for this conditional to hold water.

The bonding of a conditional to its enthymematic basis is such that one of the effective ways of classifying such conditionals is by the subject matter at issue. Thus consider:

If an integer is prime, then it will not be divisible by 4.

The enthymematic basis of this conditional is simply the definition of a "prime number" as an integer divisible by no other apart from 1. And of course any number divisible by 4 will also have 2 as divisor. This circumstance makes the conditional at issue an *arithmetic* one.

It is worth noting in this regard that a conditional can provide the enthymematic basis for others. Thus consider the conditional:

Only if switches no. 1 and no. 2 are both thrown will the engine start.

Clearly this conditional itself can serve as enthymematic basis for the following conditional:

If switch no. 1 is not thrown, then the engine will not start.

For the initial conditional has the format:

(1) $\sim$(no. 1 & no. 2) $\Rightarrow$ $\sim$*Start*

or equivalently:

Start $\Rightarrow$ (no. 1 & no. 2)

But now consider the hypothesis:

(2) $\sim$no. 1, by assumption

Then clearly:

(3) $\sim$*Start*, from (1) and (2).

Since (3) is deducible from (1) and (2), we can obtain the conditional (2) $\Rightarrow$ (3), with (1) hovering in the background as its enthymematic basis. Accordingly, that original conditional here serves as enthymematic basis for: "If switch no. 1 is not thrown, the engine will not start."

4   Iffy Variations

Conditionals are often disguised so that any explicit mention of "if-then" is absent. Thus "He would have been a fool to do that" is to all intents and purposes equivalent to:

If he did that, he was a fool.

And even with explicitly iffy statements there can be many variations.

Even If

What role does "even" play in the conditional: "Even if p, then q" (e.g., even if he had sworn it on the Bible, I would [still] not have believed him)? It is an indication of counterexpectation, operative in a context where one would normally expect that if p, then not q. An even-qualified conditional counterindicates an otherwise natural expectation in the case at hand. Thus consider: "Even if they paid him a lot more, he would (still) be disaffected." Normally one would expect: "If they pay him a lot more, he will not be disaffected." Or again consider:

Even if you were the last man on earth, I (still) wouldn't marry you.

Ordinarily one would expect a healthy young female to show some interest in partnering with the only available young male, but our conditional indicates a personal antipathy that demolishes this general expectation in the present case. (Note the role of "still" as a further reinforcement.)

On this basis, when there is an auxiliary power generator in place, it would be appropriate to say:

Even if the main cable breaks, electric power will still be available.

And an analogous situation arises in all cases of redundant causality and of fail-safe or backup provision, as per:

Even if the gunshot had not killed him that morning, the poison he had ingested at breakfast would have.

Even if assassin no. 1 had failed, assassin no. 2 would have done him in.

On the other hand, consider a counterfactual conditional on the order of:

Even if Caesar had not crossed the Rubicon, Washington might still have crossed the Delaware.

Here we have a counterfactual conditional whose consequent is a possibility and whose antecedent concerns a fact that is substantially independent of it. Such conditionals are invariably tenable—albeit not very informative, since the antecedent does little real work.

Only If

"Only if p then q" generally implies "If q then p." Thus

Only if one is in New England will one be in Vermont

will imply

If one is in Vermont, then one will be in New England.

However, there are exceptions to this rule. Thus,

Only if he swears to it will she believe what he says

does not unproblematically imply

If she believes what he says then he swears to it.[6]

For it is a tacit aspect of that original conditional that its "only if" clause bears the unspoken qualifier "among the things she does not otherwise accept but regarding what relies entirely on his say-so." But this tacit qualification does not automatically accompany the second conditional.

If Only

An if-only conditional on the order of

If only he had known that the assassins were lying in wait, he would have avoided going there

comes to much the same thing without the "only," whose sole role here is one of stylistic emphasis on the positive aspect of the consequent. Thus consider a statement on the order of:

If only he had explained, she would have forgiven him.

This conjoins an ordinary if-then conditional with an evaluative indication that realization of the consequent is a good thing. A paradigm instance of this sort of thing is:

If only he would give up drinking, he would be a far happier person.

This example shows that if-only conditionals need not be counterfactual. They can be anticipatory with an aura of qualified hopefulness.

If and Only If (IFF)

A conditional of the format "Iff p, then q," comes to: If p then q, and if q then p. This is generally called a *biconditional* because its conditionality runs both ways. What is generally at issue is the formulation of a necessary and sufficient condition as per:

If and only if he studies will he pass the examination.

In ordinary usage, "if" is often employed to mean "if and only if" (iff). When the gatekeeper says "You may enter if you have a ticket," his stipulation is meant to convey not just an if but also an only if.[7]

As If

Consider the conditional: p as if q ("He greeted her cordially [p] as if—or 'just as if' or 'as though'—they had not quarreled [q]"). This is a close cousin of the "even if" locution. We effectively have a combination of an ordinary conditional with an index of present-case violation:

Ordinarily: If p, then q (and thereby: If not q, then not p).

In the present case: p, despite not q.

Thus the case at hand runs against the general rule, by way of exception. Consider, for instance: "He threw his weight about, as if he were the top dog," combines

If he were not the top dog, then (ordinarily) he would not throw his weight about

with

In the present case: he threw his weight about even though he was not the top dog.

The as-if locution marks the present case as an exception to the situation usually prevailing with the conditional relationship at issue.[8]

Provided That and Unless

Some theorists maintain that "provided that" encodes not just a conditional but a biconditional.[9] This seems doubtful. Thus suppose that Tuesdays and Thursdays are market days. You and I now discuss whether the market is open today, being in agreement that it is either Thursday or Friday, but disagreeing as to which. In the circumstances you would unproblematically say: "The market is open, provided it is Thursday." But if you were to say "The market is open if, but only if it is Thursday" your statement would seem to deny (erroneously) that Tuesday is also a market day. Here the two assertions are not equivalent.

And the same goes for "unless," theorists to the contrary not withstanding.[10] In saying

You won't pass the exam unless you study,

I am effectively claiming

If you don't study, then you won't pass.

But I am offering no definite assurance that if you do study you will pass. It is a necessary but not necessarily sufficient condition that is at issue.

In general, it should be remarked that accepting "If p, then q" is something very different from accepting "q on condition that p" or accepting "q provided that p." Conditional or provisional acceptance is a matter of

(1) $p \Rightarrow$ to accept q (here "... $\Rightarrow$ ____" is to be read as "If ..., then ____")

and not

(2)   to accept $(p \Rightarrow q)$.

For with (2) we do actually accept something here and now, namely, that a certain implicative relationship ($\Rightarrow$) holds between p and q. With (1) by contrast we do no more than to indicate the acceptability of something (viz., q) if and when a certain condition (viz., p) is realized—which may well never occur.

These considerations do not, however, gainsay the fact that "unless" and "provided that" are frequently used in the sense of "if and only if." The long and short of it is that "unless" and "provided that" admit of different constructions in different situations.[11]

5   Uses of Conditionals

Often conditionals are used even for straightforwardly factual reportage. Thus to say that the town hall is three blocks ahead and one to the left one could offer the conditional:

If you go three blocks ahead and one to the left you will be at the town hall.

The fact is that whatever information can be conveyed in a declarative statement can be formulated in conditional form as well. Thus consider a statement p—or equivalently "p is true." Observe that precisely the same information is conveyed by the conditional:

If you answer "yes," then you are [or: *will be*] giving the correct reply to the question: "Is p true or not?"

There is no difference in informative content here: what difference there is is purely stylistic. Of course, this very fact shows that the aspect of conditionality is quite inessential in this particular case and provides merely a stylistic variation.

Disposition statements are always tantamount to conditional if-thens. ("Glass is fragile" amounts to "If glass is subjected to a shock, it will break"; "Wood is combustible" comes to "If wood is exposed to fire, it will burn"; and so on.) And many statements have conditional implications and involvements. "I must meet him lest he complain" implies "If I don't greet him, then he will complain." All sorts of capacities and capabilities are conveyed by if-then statements. "He knows how to do that sum" comes to "If you ask him to do that sum, he will provide the correct answers." "If one plays chess with him he will acquit himself ably" comes to "He plays chess well." Moreover, all sorts of conditional implications are inherent in the fact that qualifying as something of a given kind inheres in keeping true to type. If it does not look like a duck, waddle like a duck, and quack like a duck, we would be reluctant to call it one.

Conditionals are especially prominent in causal contexts, and conditional and causal talk are often interconnected. However, it is important to note in this connection that "p because q" admits of two constructions: the rationalizing and the causal. Some examples help to clarify this.

Rationalizing (Rational Explanation)

He is in France because he is in Paris.

$X > 3$ because $X > 4$.

In such a case it is not that being in Paris *causes* being in France or that being greater than four *causes* being greater than 3. It is rather that our entitlement to say the one thing provides for an intellect to say the other.

The stipulative use of conditionalization proceeds on the order of "If you do the work, I'll pay you" or "If he goes, we shall join him." The causal type has such instances as "If the match is struck, it will light." The performatory type involves performance conditions and includes conditional orders, promises, intentions, and the like. ("I'll do it, if he orders me to.") The inferential type maintains implicational consequences. ("If people are in Paris, then they are in France.") Evidential conditionals have such instances as "If that's what he said, then he must have been thinking about her."

In sum, conditional statements are ubiquitous and serve a wide variety of communicative functions:

Conjecture   If there were life on other planets, it might be scientifically more advanced than we.

Explanation   If he left, she must have been arguing with him.

Causality   If cancer could be cured, life expectancy would increase enormously.

Process description   When that switch is closed, the light goes on.

Planning   If tomorrow is a nice day, we'll go on a picnic.

Intention avowal   If I remember it, I will greet him by name.

Habits and dispositions   If addressed in the early morning, he was curt.

Instruction   If you want to start the engine, just turn the key.

Control   If that handle is turned clockwise, water will flow from the spigot.

Promises   If you require the money, I'll pay it back to you.

Contracts   I'll do *A* if (and only if) you'll do *B*.

Excuses   If only I had known how sensitive he is, I would never have said that.

Restrictive requirements   "If a child attends, it must be accompanied by a parent" reformulates "Children who attend must be accompanied by parents"; "If a building has not passed inspection, then it cannot be insured" comes to "Only buildings that have passed inspection can be insured."

Threats   If you do that, we'll make sure that you do no more business in this town.[12]

Negotiation   Only if you do *A* will we do *B*.

As even so brief and incomplete a survey indicates, conditionals are an enormously versatile resource.

What would be lost by doing without conditionals—and so without hypothetical reasoning at large? The very question is an exercise in self-illustration. We would be unable to speculate, to plan, to make inference, and in general to achieve a vast variety of communicative functions. The limits of our thought would contract to the limits of our factual knowledge, confined to what we accept as matters of real truth. Speculation, conjecture, fiction would all be outside our repertoire. Even science itself, which standardly proceeds by way of hypotheses and thought experiments, would become incapacitated. Without the use of conditionals speculative thought would be infeasible: we could not deal efficiently with questions of "what-when?" if the resources of "what-if?" were not at our disposal.

2   Matters of Aspect

1 Aspect

Conditionals, just like unconditional statements, are subject to different sorts of grammatical qualification: time and tense, modality, mood, adverbial mode, and the like. And various of their characteristic logico-grammatical features emerge in these regards. We shall here refer to these logico-grammatical features of statements generically as *aspects*, borrowing this term from its more specialized employment by grammarians.

2 The Cognitive Aspect

There are three significantly different ways in which the antecedent (protasis) p of a conditional $p \Rightarrow q$ can be related to the body B of relevant background knowledge—it can be believed, disbelieved, or belief-suspended:

(1) $p \in B$: p is known or accepted as true, in which case we have a "since conditional" rather than an iffy one: "Since p is true, q is."

(2) $\sim p \in B$: p is known or accepted as false: "If p were true, q would be."

(3)   Neither $p \in B$ nor $\sim p \in B$: p is of unknown truth status: "If p is true, q will be."

These three cases present us with *factual, counterfactual,* and *agnostic* conditionals, respectively.

In case (1), the (antecedent) supposition itself already belongs to our background knowledge. And in case (3) the supposition can in principle simply be added to our background knowledge. (The supposition has to be compatible with *B* because if it weren't it would be deemed false.) However, from a theoretical point of view, case (2) is far and away the most interesting, exactly because it raises the problem of how a belief deemed to be false is to be reconciled with the wider framework of beliefs with which it is incompatible. (The difficulties that arise here were already noted by the medieval scholastics—as we shall see in due course.)

The simplest case here is that of *factual* conditionals, whose antecedents we accept as true. Here there is no "if" or "but" with respect to an "assumption"—exactly because the antecedent is accepted as true. In ordinary usage these factual conditionals are often given a distinctive formulation by using "since" in place of "if-then."[1] Thus when we know that John is in Paris, we would ordinarily say not:

If John is in Paris, then he is in France,

but rather:

Since John is in Paris, he is in France.

Only the former of these is genuinely conditional—and *agnostic*. Factual "conditionals" of the latter sort are so in name (and form) only: there is nothing really iffy about them for anyone who knows that Paris is in France.

Factual conditionals stand in diametrical contrast to *counterfactual* or *belief-contradicting* conditionals, whose antecedents we regard as false.

If George Washington had been a foot taller, then his height would have exceeded seven feet.

If F.D.R.'s image were added to the Mt. Rushmore sculptures, then five former presidents would be depicted there.

Counterfactual conditionals are usually characterized as such in English by the grammatically conditional formation of a past-tense antecedent (were, had been) followed by a conditional marker (would, could, might) plus use of the subjunctive in the apodosis. Thus, if we happen to know that John is in London, we would say not:

If John is in Paris, then he is in France,

but rather:

If John were in Paris, then he would be in France.

Sometimes the distinction between factual and counterfactual conditionals is represented as faithfully mirroring that between indicative and subjunctive conditionals. But while such a correspondence holds in many and indeed most cases, it is not always so. The indicative conditional

If he invites me (which he won't), I will certainly refuse

is every bit as much counterfactual as the subjunctive

If he were to invite me, I would certainly refuse.

By contrast, the subjunctive conditional

If you were to check the text of Shakespeare's *Twelfth Night*, you would find that he speaks of *midsummer* (and not *midwinter*) madness

is every bit as factual as the indicative

If you were to check the text of Shakespeare's *Twelfth Night*, you will find that he speaks of *midsummer* (not midwinter) madness.

Note that in both cases one could unproblematically attach the rider "and I expect you will" to the antecedent.

Conditionals whose antecedent is acknowledged as false represent a special and distinctively complicated issue. Examples are given by such counterfactual conditionals as:

If wishes were horses, then beggars would ride.

If Caesar had heeded the seeress, he would not have proceeded to the senate.

Providing an adequate account of counterfactual conditionals is one of the principal challenges for a theory of conditionals because of the difficulties created by a conflict with that which we actually know or believe.

Intermediate between the factual and the counterfactual cases there are *speculative* or *agnostic* conditionals, regarding whose antecedents we are undecided, neither believing nor disbelieving them. Scheduling information affords a good example of agnostic conditionals:

If the boat leaves on time, it will reach its destination on the other side of the lake at 3:00 PM.

This is based on how things should go normally, but we have no idea whether or not it will be so on *this* occasion.

Despite the unknown status of the antecedent, such conditionals will avoid the subjunctive:

If Charles Dickens did meet Queen Victoria (and I don't know if he did or didn't), then he must have been impressed.

If the president is currently enroute to China, then he will not be at Camp David this weekend.

As such examples indicate, one's cognitive stance toward the antecedent condition at issue determines how one would properly formulate a conditional. But the rules of procedure are not altogether cast-iron. In the past, speculative conditionals were usually formulated with a subjunctive antecedent and a grammatically *conditional* consequent:

If he were to come [which is unknown one way or the other], then we would have a foursome.

If it were to rain [which is something we don't know one way or the other], then the newly planted seed would be washed away.

However, modern usage—which tends to minimize use of the subjunctive—often formulates such conditionals in the future indicative:

If he comes, we'll have a foursome.

If it rains, the seed will wash away.

And so, today we would not say the Shakespearean

If it were done, then 'twere well it were done quickly.

but rather, avoiding the subjunctive, would say something like:

If it is going to be done, then it had best be done quickly.

The tabulation of table 2.1 summarizes how the proper formulation of conditionals hinges on the cognitive status of the antecedent and consequent, with the cognitive status of a conditional's components mirrored in the variation of their mood and tense.

As these considerations indicate, the belief-geared cognitive status—or, as some writers prefer, *epistemic* or *doxastic* status—of its antecedent is pivotal for determining the formulation and thereby the meaning of a conditional statement.

Table 2.1
Conditionals and Epistemic Validation

Antecedent seen as	Consequent seen as	Formulation of: (antecedent) → (consequent)
T	T	Since that bird is a canary, it is yellow.
T	F	Case excluded
T	?	Case excluded
F	T	If that bird were a canary, it would (still) be yellow.
F	F	If that bird were a canary, it would (have to) be yellow.
F	?	If that bird were a canary, it would be yellow.
?	T	If that bird is a canary, it will (also) be yellow.
?	F	Case excluded
?	?	It that bird is a canary, it will be yellow.

KEY: T = deemed to be true
F = deemed to be false
? = of unknown truth status
Note: If the consequent is seen as false, then a valid conditional will have to be a counterfactual whose antecedent is also seen as false. And if the antecedent is seen as true, then the consequent cannot be other than true. The consequent of an appropriate conditional cannot possibly be weaker than the antecedent in the T, ?, F order.

3 Hypotheticality

Consider the following series of conditionals:

When he comes to the conference tomorrow, I will finally get to meet him.

If he comes to the conference tomorrow, I will finally get to meet him.

If he came to the conference tomorrow, I would finally get to meet him.

If he were to come to the conference tomorrow, I would finally get to meet him.

While these convey much the same information about the relationship between the subject's conference attendance tomorrow and my (finally) getting to meet him, there is a significant difference between these conditionals: they put the antecedent-contemplated issue of his conference attendance tomorrow into an increasingly doubtful and problematic light. In that first "when" conditional, the antecedent's eventual realization is seen as virtually certain. Thus, with each successive step, there is a diminution of the estimated likelihood of the antecedent's realization, so that the degree of hypotheticality is continually intensified. With (1) there is a

Table 2.2
Differential Assessments

Commitment	Assessed Hypotheticality	Likelihood
• acceptance	• reality	• certainty
• positively inclined uncertainty	• real possibility	• high probability
• negatively inclined uncertainty	• remote possibility	• low probability
• total rejection	• impossibility	• zero probability

tacit "where I deem certain". But by the time we get to (4) there is a tacit "which I have no idea whether or not it will happen". There seems to be a scale of distinct epistemic levels of hypotheticality and/or assessed likelihood, as depicted in table 2.2. And these levels of hypotheticality are reflected grammatically in the way the conditionals are formulated.

It must be noted, however, that the expectation-confident use of "when" differs from its no less common use of "whenever." Thus in telling you that "When in Rome, do as the Romans do" I do not mean to imply confidence in your eventual presence in Rome. And saying "When inflation comes, prices rise" or "When in doubt, he punts" issues no assurance that the antecedent circumstance will be realized, but rather uses "when" in the sense of "whenever" to effect a generalization.

Suppose that we know (as part of the accepted enthymematic basis for conditionalization) that Tom answers the doorbell whenever he is home (and of course only then). There are now four possibilities:

1. The status of the antecedent is settled—we know Tom is (or isn't) home:

Since Tom is [is not] home, he will [will not] answer the bell.

2. The status of the antecedent is open—we don't know/believe one way or the other that Tom is home:

If Tom is [or: *will be*] home, he will answer the bell.

3. The status of the antecedent is doubtful—we are inclined to think it false:

If Tom were home, he would answer the doorbell.

4. The status of the antecedent is negative: we know/believe it is false:

If Tom had been home, he would have answered the doorbell.

We thus have a sliding scale from factual to counterfactual (belief-contravening) conditionals.

In English, the prospect of a conditional's realization is frequently indicated by tense:

Neutral: *present + future*: "If he comes, he will see me."

Unlikely: *subjunctive or indicative past + conditional*: "If he came, he would see me"; "If he should [or were to] come, he would see me."

Outright false: *pluperfect past + conditional past*: "If he had come, he would have seen me."

As this indicates, the stronger the shift to the past, the more unlikely the realization at issue is held to be.

Use of the past tense in the antecedent often serves to indicate hypotheticality not only in the manner of improbability but also in the manner of unreality. In saying things on the order of

If wishes were horses, beggars would ride

If I were you, I would kill him

what one has in view is not just that the antecedent is improbable, but rather that it is totally unrealistic. (To be sure, even present-tense conditionals can have manifestly unrealizable antecedents: "If 6 is prime, then I'm a monkey's uncle.")

4 The Temporal Aspect

Compare the following:

After I finish my work, I will go home.

When I finish my work, I will go home.

If I finish my work, I will go home.

The transition from temporal posteriority to conditional consequentiality is smooth and seamless.

Moreover, in such contexts the role that "if" plays is effectively adverbial: "If I go, you go" has a format akin to "Where you go, I go," "When you go, I go," and "After you go, I go." All have substantially the same structure:

In any circumstance: If it is such that you go, then I go as well.

With whatever place: If it is where you go, then I will go there as well.

At whatever time: If you go at that time, the I will go at that time as well.

At whatever time, if you go at that time, then shortly after I will go as well.

The temporal–atemporal distinction bears importantly on the explanation of conditionals.[2] Atemporality is instanced by such conditionals as:

If $x > 3$ then $x > 2$.

If someone is at the North Pole, then every direction in which he moves leads southward.

With such conditionals the factor of time is substantially irrelevant.

By contrast, temporal conditionals proceed with respect to the distinction of past, present, future, and omnitemporal, some of which admit of further subdivision.

Present

If that is a moose, then it is a very large one.

If that is he at the door, they will permit his entry.

Future

Expected   When he comes, they will greet him.

Speculative (and thus expectation-neutral)   If he comes, they will greet him.

Counterexpected   If he were to come, they would greet him.

Past

Episodically actual   When he came, they greeted him. [Strictly speaking, there is no conditionality here but simply sequential conjunction: he came; they greeted him.]

Habitually factual   Whenever he came, they greeted him.

Agnostic   If he came, they (doubtless) greeted him.

Counterfactual   If he had [or: *would have*] come, they would have greeted him.

The temporal aspect of a conditional is determined by its meaning and not just by its grammar. Thus

If he comes, they are likely to be upset

is both future and speculative in status, even though both antecedent and consequent are present-tense indicatives.

And there is yet another temporal category.

Omnitemporal

If [or: *whenever*] it rains, it pours.

If [or: *whenever*] you boil water, it bubbles.

The omnitemporal aspect of the conditional is indicated by the circumstances that we could add "ever" in the antecedent without changing the meaning at issue.

The temporal classification of conditionals cuts across the factual–counterfactual divide.

Factual

Present   If he does that, he is a fool.

Future   If he does that, he will be a fool.

Past   If he did that, he was a fool.

Counterfactual

Present   If he were to do that, he would be a fool.

Future   If he were to do that, he will be a fool.

Past   If he had done that, he would have been a fool.

It should be noted that a temporal readjustment may be necessary to accommodate the meaning of conditionals in the face of their logical reformulation. Consider, for example, the case of contraposition of a conditional from $p \Rightarrow q$ to $\sim q \Rightarrow \sim p$. With the causal or productive if-then, whose meaning in essence is

If p, then (as a result) q,

one clearly needs to make a suitable change of tense with contraposition. For example:

When he overeats, then his stomach will (subsequently) ache

needs to be contraposed as:

When his stomach does not ache, then he has not (previously) overeaten.

As such examples show, for contraposition to work out properly a conditional's tensing may have to be revised to keep the time sequence straight.[3]

5 Time–Cognition Interaction

In general, formulating conditionals in English is a vastly complex matter, and whatever rules there are are guidelines rather than fixed rigidities. Often it depends not on the time of action but on when information about the action becomes available. When it is said that

If he passed the exam [which was taken last week and graded yesterday], it would be a miracle.

it is because information about the result yet awaits the future, while the fact itself—passed or failed—is already settled. The issue of temporality in such matters is complex and the differences of meaning that result are sometimes subtle.[4]

Consider the following array of antecedent-agnostic conditionals that arise when we do not know whether or not someone is lying to someone else, but we do know that their relationship will be greatly at risk if he deceives her:

Present tense   "If he were lying to her, he would be running [or: *would run*] a great risk"—or, in a subjunctive-avoiding format: "If he is lying to her, he runs [or: *is running*] a great risk."

Future tense   "If he lies to her, he will run [or: *will be running*] a great risk." (This could also be put "If he lies to her, he runs a great risk"—with a possible present-time applicability.)

Past tense   "If he lied [or: *has lied*] to her, he ran a great risk."

Observe that these conditionals are not belief-contradicting but simply agnostic. By contrast, if one accepts the antecedent as true, one should change that initial "if" to "since" and adapt the tenses accordingly.

Present tense "Since he is lying to her, he is running [or: *runs*] a great risk."

Future tense "Since he will lie to her, he will run [or: *will be running*] a great risk."

Past tense "Since he lied [or: *did lie*] to her, he ran a great risk."

However, if we propose to reject the antecedent as false, and thus regard the conditionals as counterfactual, then we will have to say:

Present tense "If he were lying to her, which he isn't, he would be running a great risk." (Note: the rejection of the antecedent should ideally be made explicit here, as is also the case with the future tense.)

Future tense "If he lies to her, which he won't, he will run [or: *will be running*] a great risk.

Past tense "If he had lied to her, he would have been running a great risk." (Note: only here is that parenthetical antecedent-rejection readily dispensable.)

An additional complication arises in the past-oriented cases when the consequent of a conditional is also accepted as true. In this case careful usage would then have us introduce a marker such as "still" or "anyway":

If he had answered that question correctly, he would (still) have failed the exam (anyway).

6 Generality

Generality is yet another important aspect of conditional statements. Consider such statements as:

If it rains, it pours.

If you are in Paris, you are in France.

If an integer is prime, it cannot be divided by 4.

The "if" that is operative here can effectively be replaced by "whenever" or "in all instances when."

Of course, conditionals do not always involve this sort of generality: a single event or situation is often at issue.

If you apologize for that remark, she will forgive you.

If he arrives on time, then we'll take him along to a movie.

Moreover, there are also mixed cases—generalizations that hold for a particular period only:

If you order this month, you will receive a 10 percent discount.

7 Causality

Suppose that Pavlovian conditioning has established the general causal relationship that my dog barks whenever I sing. The following conditionals will then all hold:

Factually transtemporal   If [or: *when*] I sing, my dog barks.

Speculatively future oriented   If I were to sing, my dog would bark.

Counterfactually past oriented   If I had sung, my dog would have barked.

One and the same temporally stable causal linkage will thus give rise to conditionals of very different aspect, and thus one and the same enthymematic basis once again provides the grounding for a variety of conditionals.

Causal conditionals generally envision a temporal sequence of natural succession with the antecedent temporally preceding the consequent:

If you light the fire, then the room will heat up.

But the time order can also be reversed, if we look to causal prerequisites rather than consequences, so that the consequence-circumstances precede those of the antecedent:

If he is to succeed, then he must work hard.

If the water is flowing, then the spigot is [or: *must be*] open.

If the ground is wet, then it has rained.

Accordingly, the "if-then" of causal connection need not always be temporally sequential: antecedent first, consequent later. And it is certainly something quite different from a merely truth-connective material conditional.

However, while various conditionals are based on causal relationships, many are not. Thus, consider:

If tomorrow is Monday, then today must be Sunday.

This is clearly not causal: its being Sunday today is not *caused* by its being Monday tomorrow. (For one thing, causality does not flow backward.) The enthymematic basis on which this conditional rests is simply the sequential order of the days of the week, that is, that Monday always follows Sunday and that Sunday always precedes Monday.

Again, consider:

If someone is in Paris, then he is in France.

Here too there is no question of causality. Location in one domain does not *cause* location in an encompassing one. The enthymematic basis of this conditional is simply the politico-geographic fact that the city of Paris is located within the country of France. There is clearly no causality at issue here.

8 The Adverbial Aspect and Modalities

Adverbially qualified conditionals arise when we have not just plain "If *p*, then *q*," but also such accompaniments as:

- an index of *modality*: possibly, necessarily, standardly, normally;
- an index of *temporality*: yesterday, last year, henceforward;
- an index of *frequency*: always, usually, often, sometimes;
- an index of *evaluation*: mercifully, unfortunately; or
- an index of *rejection*: not, it is not the case that.

In fact, all of the ways in which propositions-in-general admit of adverbial qualification will also apply to conditional propositions.

Adverbial qualifiers can attach either to a conditional as a whole or just to its consequent. The conditional

Fortunately, if we perform this test, then we can detect a cancer (Or: If we perform this test, then, fortunately, we can detect a cancer)

says something evaluative about the *relationship* at issue between testing and diagnosis. In this respect it differs from

If we perform this test, then we can fortunately detect a cancer.

Here those parenthetical commas make a big difference.

With conditionals that are adverbially qualified (say by @) we must accordingly distinguish between modes of qualification that are *consequent* (or *internal*) and those that are *consequential* (or *external*), that is, between

@ (If p, then q)

and

If p, then @q.[5]

A significant conceptual difference is at issue here. Take, for example, the qualification *inevitable*. From $3 > 2$, it is inevitable that if the number of people in the room is now >3, then it is >2; but it is certainly not the case that if the number of people in the room is now >3, then the number of people in the room is inevitably >2. Again, it is unproblematically true that

Obviously: If p, then p,

whereas it is far from true that:

If p, then obviously p.

Life would be simpler than it is if, claims that are true would thereby also be obvious.

Again, consider rejection. Clearly the negation of the conditional as a whole as per

Not: If p then q

is something quite different from its assertion with a denied consequent:

If p, then not q.

This becomes evident when we set the variables as follows:

$p = X$ is an adult

$q = X$ is male

The difference between consequent and consequential adverbial qualification is particularly significant in the case of a modality such as possibility or necessity. Consider necessity. There is a big difference between (1) $p \to \Box q$ and (2) $\Box(p \to q)$. This is particularly clear when $p = q$. For then the second is (trivially) true but the first emphatically not, since it is patently false that every true proposition is necessary. With (1) we have a conditioinalized necessity: necessity under a specified condition. "If a number is divisible by 2, then it is necessarily even." With (2), on the other hand, we have a necessitated conditional, a necessary inferential relationship: "Necessarily, if he is older than 5 then he is older than 3." (Note that here there is nothing necessary about his age as such.) An analogous situation obtains with respect to possibility—or any other sort of modality, such as the evaluative "It is a good thing that." For a consequential modality qualifies the conditional as a whole, while a consequent modality qualifies only the consequent. For example, consider the case of *possibly* or *perhaps*. "If I pick an integer at random, it will perhaps be 3" is clearly true. But "Perhaps: 'If I pick an integer at random, it will be 3' " is clearly false. That statement is simply wrong, there is no "perhaps" about it.

The difference between internal (consequent-oriented) and external (consequential) qualification is highly important in situations of inferential reasoning. The following inference is clearly valid:

p
If p, then $F(q)$
$\therefore F(q)$

But it need by no means transpire that we will have:

p
$F(\text{If } p, \text{ then } q)$
$\therefore F(q)$

To see that *modus ponens* does not work in this second, consequential case, consider once more the modality of necessity:

Jane is a brunette.
Necessarily: If Jane is a brunette, then she is not a blonde.
$\therefore$ Necessarily: Jane is not a blonde.

Despite the *ex hypothesi* truth of the premises, the conclusion here is clearly false.

Table 2.3
Relativized Modalities

	Internal	External	Relational
Necessity	$p \to \Box q$	$\Box(p \to q)$	$N(q/p)$
Possibility	$p \to \Diamond q$	$\Diamond(p \to q)$	$P(q/p)$
Probability	$p \to Lq$	$L(p \to q)$	$L(q/p)$

Note: $L(q/p)$ could be construed as "q obtains in most cases when p does."

In view of this situation, all those aspects characterized as *modalities*—necessity ($\Box p$), possibility ($\Diamond p$), probability (Lp), and so on—always admit of two versions, absolute and conditionalized.[6] We accordingly arrive at the situation summarized in table 2.3. It is important to note in this connection that conditionalized modality on the order of "If p, then necessarily q" will always admit of the three different versions illustrated in table 2.3. And these three versions of relativized modality will in general say very different things, as will become clearer in subsequent discussion. Only in very special cases will any two of them coincide. (One such exception is that of the semantical modality "is true," when it transpires that all three versions come to the same thing.)

3 Modes of Implication

1 Deducibility

Deducibility is clearly a route to conditionals. Whenever q is derivable from p by logico-conceptual means—that is, when $p \vdash q$ obtains—then the claim "If p, then q" is clearly in order. But what sort of "logic" is at work with the deducibility at issue here?

A "logic" is a systematic codification of the rules that govern the inferential practices of a community that has occasion for exact thinking. Just as (traditional) grammar is a systematization of the linguistic practice of "grammatically careful speakers," so logic is a systematization of the inferential practice of "logically careful thinkers." (That may sound circular but what is involved here is not the vicious circularity of definitional question-begging but the virtuous circularity of an explanatory feedback loop.)

In the setting of the present discussion we shall speak simply of "logic" and of "logical deducibility" (as here represented by $\vdash$). Yet it must be acknowledged that this is something of an oversimplification. Nowadays there of course exists a wide variety of different sorts of logical systems, not just classical logic alone, but intuitionistic logic, many-valued logic, fuzzy logic, existentially "free" logic, and many others.[1] What, then, is the appropriate way to react to this prospective proliferation of logical systems?

There are three prime alternatives here:

1. *Pluralism* takes the following stance: You are free to pick your own favorite style of logic. And after you have done so, and the deducibility relationship $\vdash$ is thus fixed, then the rest of the analysis of conditionals proceeds from that point on to yield a choice-relative result. Deducibility,

in short, hinges on what the favored logic prescribes. And there will, inevitably, be alternatives here as between different schools of thought.

2. *Canonicalism* proceeds on the basis of instituting one particular system of logic—presumably classical, truth-functional logic—as the standard, canonical logic on which deducibility is to be based. Other alternatives are dismissed from the range of concern.

3. *Dominance* construes deducibility in a synoptically comprehensive way by taking the line that the generalized deducibility relation $\vdash$ holds whenever there are no instances in *any* particular logic in which the corresponding relation authorizes a false conclusion from true premises.

In principle, any one of these alternatives can be adopted. Each of them is inherently available as a practicable mode of operation, and each will issue in a viable theory of conditionals. One merely needs to recognize and accept that one's understanding of if-then conditionalization will be geared to the approach one adopts. For the long and short of it is that the acceptability of conditionals is inseparably linked to the operations of inferential logic. And this means, among other things, that insofar as people differ regarding logic they are bound to differ regarding the acceptability of conditionals as well.

For present purposes, however, we shall adopt what seems to be the most straightforward alternative, namely to accept classical deductive logic as canonical.

2   Understanding Conditionals in Enthymematic Terms

As noted above, the propositional if-then conditionals that primarily concern us here have two principal components, an *antecedent* (protasis), which states the condition at issue, and a *consequent* (apodosis), which states the result held to ensue from the realization of this condition. We shall here refer to "if-then" ($\Rightarrow$) relationships in general as conditionalization, contrasting this generic idea with such more specific relations as logico-conceptual deducibility or derivability ($\vdash$) on the one hand, and on the other the relationship of implication ($\rightarrow$), which are defined in terms of obedience to basic logical rules that are yet to be specified.

If a conditional is to assert a tenable relationship, then its consequent should ideally be related to and connected with the antecedent in such a way that it can be said to "follow" from it in some plausible sense of that term. But while deductive derivability is sufficient to validate conditionality, it is certainly not necessary. A broader construal along indirect lines will also meet the needs of the situation by having it that the antecedent yields the consequent as a deductive consequence in the presence of a suitable body of background information. Thus, for example, the agnostic conditional

If that spoon is made of silver [and we don't know this one way or the other], then it will conduct electricity

is validated by the fact that the conjunction of the antecedent ("That spoon is made of silver") together with some suitable item of background information ("Silver conducts electricity") makes it possible to deduce the conclusion by logic alone.

The background information that is needed to validate a conditional relationship may be designated as the *enthymematic basis* that provides the linkage underlying the conditional at issue. For example, the counterfactual conditional

If $\frac{1}{2}$ were an integer, then it would be either odd or even

has as its enthymematic basis that every integer is either odd or even.

Again, consider the following scenario. The Merchant's Bank suffered an armed robbery last week. Smith is accused of the robbery. He claims that he was out of the country. Now even if we are highly skeptical about his claim, we will nevertheless have little hesitancy about endorsing:

If Smith was out of the country last week, then he did not commit the robbery in question.

We accept this contention because of our general realization that to commit an armed bank robbery one has to be at the scene. When this background information is added as "enthymematic basis" to the antecedent of this conditional, the consequent follows as a simple matter of logic.

Seen from this point of view, a conditional effectively summarizes an enthymematic argument.[2] On this approach an if-then statement of the format:

antecedent $\Rightarrow$ consequent

is, in effect, the abbreviated report of a deductive argument of the form:

(antecedent + enthymematic premisses) $\vdash$ consequent

For example, the conditional

If you agitate him, he will probably stutter

should be seen as based on the statistical fact that he generally stutters when agitated.

According to the Oxford philosopher F. H. Bradley, a supposition (or "supposal" as he prefers to call it)

is an ideal experiment. It is an application of a [suppositional] content to the real, with a view to seeing what the consequence is, and with a tacit reservation that no actual judgement has taken place. The supposed is treated as if it were real, in order to see how the real behaves when qualified in a certain manner.[3]

This way of linking supposition to reality is quite in order. Thus the supposition-based conditional "If you were in Paris, you would be in France" is bound hard and fast to the real-world situation of Paris being in France. Such a fact, or putative fact, is indispensably requisite to provide the basis for any viable supposition-based conditional. And a conditional must, when correct, rest (implicitly and tacitly, to be sure), upon such matters of fact.

Moreover, one and the same fact (or putative fact) can provide the epistemic basis for a variety of conditionals. Thus let it be that Jim is invariably accompanied by his dog, Rover. Then any of the following conditionals can appropriately be maintained:

Since Jim is present, Rover is (or: *will be, must be*) present as well.

If Jim was present, Rover was present.

If Jim had been present, Rover would have been present.

Factual, speculative, and counterfactual conditionals can all rest on the same enthymematic basis.

The tenability of a conditional is thus predicated on the existence of a connecting factual link between its (appropriately supplemented) antecedent and its consequent, a circumstance signaled by medieval logicians in employing the term *connexa* for hypothetical if-then judgments.

It is worth noting that our enthymematic-deductive account of conditionals does not constitute an explicatory analysis of conditionals that "reduces" them to nonconditional statements. After all, ⊢ itself represents a conditional of sorts. Instead, what it accomplishes from an explanatory point of view is to "reduce" conditionals in general—a category that includes many obscure and problematic cases—to a particular form of conditionality that is (at least comparatively) clear and well understood.

3   Generic Implication

For an if-then relationship ($\Rightarrow$) to qualify as a mode of implication ($\rightarrow$) certain specific requirements have to be met. Every mode of implication affords a means for the formulation of conditionals. But not every conditional is an implication.

To begin with, any implication operator ($\rightarrow$) worthy of the name should also obey certain further logical principles. For example, an implication relation—and indeed any inferentially serviceable conditional—cannot move from a true antecedent to a false conclusion. We thus have:

Truth Preservation

If $p \rightarrow q$, then $\sim(p \,\&\, \sim q)$ and hence $\sim p \vee q$.

Moreover, it must transpire that whenever q is deducible from p, then $p \rightarrow q$ necessarily obtains:

Deductivity

If $p \vdash q$, then $p \rightarrow q$, and indeed $\Box(p \rightarrow q)$.

This principle of deductivity means that we will also have:

Reflexivity

$p \rightarrow p$

Moreover, any implication relationship ($\rightarrow$) worthy of the name must permit inference by *modus ponens*. That is, if $\rightarrow$ is to qualify as a mode of implication proper, then we must have:

Implicativity or *Modus ponens*

If p and $p \rightarrow q$, then q; that is, $(p \,\&\, [p \rightarrow q]) \vdash q$.

Then too, an implication relationship should also obey the rule:

Contrapositivity

If $p \rightarrow q$, then $\sim q \rightarrow \sim p$.

And these three requirements are not yet enough. Any implication relationship will further have to satisfy:

Transitivity

If $p \rightarrow q$ and $q \rightarrow r$, then $p \rightarrow r$.

Conjunctivity

If $p \to q$ and $p \to r$, then $p \to (q \& r)$.

Disjunctivity

$(p \to q) \to ([r \vee p] \to [r \vee q])$.

Also, since $(p \& r) \vdash p$ means that $(p \& r) \to p$ we will—by virtue of transitivity—have:

Monotonicity

If $p \to q$, then $(p \& r) \to q$.

And finally, implication also requires:

Substitutivity

If $p \to q$ and $q \to p$ (that is, if $p \leftrightarrow q$), then $F(p) \leftrightarrow F(q)$, provided that F is a propositional context involving only the connectives $\&$, $\vee$, $\sim$, and $\to$.

Taken together, the preceding conditions mean that true propositions can never imply a contradiction: with any viable sort of implication $p \to q$ and $p \to \sim q$ means that $\sim p$. For by contraposition $p \to \sim q$ yields $q \to \sim p$, and by transitivity this together with $p \to q$ yields $p \to \sim p$, which means that p must be false.

All these various rules are standard to those relationships that qualify for characterization as a mode of *implication*, and any relationship worthy of this name will have to obey them.[4] Indeed, on the indicated basis we have it that any implication relation, $\to$, provides for the standard calculus of propositional logic axiomatized by Hilbert and Ackermann[5] via four axioms and two rules:

A1. $(p \vee p) \to p$

A2. $p \to (p \vee q)$

A3. $(p \vee q) \to (q \vee p)$

A4. $(p \to q) \to [(r \vee p) \to (r \vee q)]$

R1. *Modus ponens*

R2. *Substitutivity of Equivalents*

We have provided for R1 and R2 explicitly. Deductivity serves to ensure A1 through A3. And A4 is explicitly provided for.

However, since logico-conceptual demonstrability, $\vdash$, represents ordinary (or natural) deducibility and $\to$ is subject to the deductivity rule that

$p \rightarrow q$ whenever $p \vdash q$, it transpires that $\rightarrow$ will be *natural* rather than so-called *relevant* implication. For relevant implication rejects $(p \,\&\, \sim p) \rightarrow q$, while natural implication is committed to it via the following line of reasoning:

1. $\vdash (p \vee \sim p)$ (as a principle of logic)
2. $\sim q \vdash (p \vee \sim p)$ (from 1)
3. $\sim q \rightarrow (p \vee \sim p)$ (from 2 by deductivity)
4. $\sim(p \vee \sim p) \rightarrow \sim\sim q$ (from 3 by contraposition)
5. $(p \,\&\, \sim p) \rightarrow q$ (from 4)

In any event, a mode of implication ($\rightarrow$) must, to qualify as such, obey a variety of "logical" principles along the lines of contraposibility, transitivity, monotonicity, and so forth. And various sorts of conditionals fail to meet this requirement—counterfactual conditionals prominently included, as will emerge below.

4 Enthymematic or Substantive Implication

The term *enthymeme* has been used by logicians since antiquity to characterize an argument with a tacit, unexpressed premise, as per: "All men are mortal. Therefore, Socrates is mortal. [Unstated premise: Socrates is a man]." But with a small step this usage can be reshaped to yield the conditional "Since all men are mortal, Socrates [being a man] is mortal."

Enthymematic implication ($[S] \mapsto$) is accordingly a mode of conditionalization that can be defined against the background of some given self-consistent set S of "available" propositions via the specification:

$$p\,[S] \mapsto q \text{ iff } (p + S) \vdash q.$$

(It is presupposed here that the "availability" at issue is such that the logical consequences of available propositions are always available.) On this definition, one proposition enthymematically implies another (relative to the enthymematic manifold S) if the former (the antecedent), when duly conjoined with various S-members to provide an enthymematic basis, provides for the logico-conceptual deducibility of the latter (the consequent).

What we have here is a supposition-based construal of if-then conditionalization, in that $p \Rightarrow q$ is held to obtain whenever the supposition of p in the context of S-provided information makes possible the derivation of q. These conditionals obviously rest on a suitable background of unarticulated substantive information, and can accordingly be character-ized as substantive conditionals. With such substantive conditionals of the format $p\,[S]\!\!\!\mapsto q$ we have a mode of conditionalization that is defined relative to a manifold S of "available" propositions whose acceptance as true provides the enthymematic bases for our conditionals.

The enthymematic nature of many if-then conditionals has long been recognized.[6] For example, "If the butler did not poison Sir John, then the cook did it" will be acceptable only to someone who accepts a justi-factory premise along the lines of "Poisoning is done only by people who have motive and opportunity" and "Only the cook and the butler had motive and opportunity for poisoning Sir John."

It should be noted that classic inference pattern of *modus ponens*, namely

$$p$$
$$\underline{p\,[S]\!\!\!\mapsto q}$$
$$\therefore q$$

is automatically encompassed in the very meaning of enthymematic im-plication, seeing that the membership of S is to be taken as given. For when $(p + S) \vdash q$, then whenever p and S are both in hand, q will be as-sured. However, when S consists not of actual truths but merely putative ones, then *modus ponens* must be construed correspondingly. For exam-ple, let S consist simply (and only) of the false belief q and its logical con-sequences. Then

$$p\,[S]\!\!\!\mapsto (p\,\&\,q)$$

will be true despite the falsity of the consequent. Accordingly, we would have it that the *modus ponens* situation

$$p$$
$$\underline{p\,[S]\!\!\!\mapsto (p\,\&\,q)}$$
$$\therefore p\,\&\,q$$

now obtains with true premises and a false conclusion. But of course $p\,\&\,q$, albeit false, will in these circumstances be a *putative* truth for the individual who, by hypothesis, accepts q as true. And in the context of our present purpose it is this that matters.

Some further general features of this implication relationship are:

If $q \in S$, then $p\,[S]\!\!\!\mapsto q$ for any p whatsoever.

If $\vdash q$, then $p\,[S]\!\!\!\mapsto q$, for any p.

Since S is self-consistent, we can never have it that $p[S] \mapsto q$ and $p[S] \mapsto {\sim}q$ as long as p is not self-inconsistent.

To show in general that $[S] \mapsto$ qualifies as a mode of implication $(\rightarrow)$ in the manner indicated above we have to show (inter alia) that it satisfies the various conditions set out above (that is, is monotonic, contrapositive, and transitive, etc.). But this is unproblematic: all of these various requisites are readily established.

5   Further Modes of Implication

The symbolic and terminological conventions outlined in table 3.1 will be employed throughout the present book. This specified notation casts implication $(\rightarrow)$ in the role of a generic conception that encompasses various modes of if-then conditionality. All of the enumerated modes of implication can be defined in terms of $[S] \mapsto$. This can be substantiated by noting the following:

Case 2 S is a set of *demonstrable* truths (D), so that in general: $p \in S$ if $\vdash p$. Then $[S] \mapsto$ is deductive entailment $\vdash$. So here $S = D = \{p: \vdash p\}$ will do, as will any subset thereof.

Case 3 S is the set of truths at large (T), so that in general: $p \in S$ iff p. Then $[S] \mapsto$ is material implication $\supset$. So here $S = T = \{p: p$ is true$\}$.

Table 3.1
Modes of Implication

Symbol	Name	Use	Reading
0. $\rightarrow$	implication (generic)	$p \rightarrow q$	"p implies q"
1. $[S] \mapsto$	enthymematic implication	$p[S] \mapsto q$ iff $(p + S) \vdash q$	"p enthymematically yields q, relative to S"
2. $\vdash$	deductive implication	$p \vdash q$	"p yields (or: deductively entails) q"
3. $\supset$	material implication	$p \supset q$ iff ${\sim}p \lor q$	"p materially implies or: m-implies) q"
4. $\prec$	strict implication	$p \prec q$ iff $\Box(p \supset q)$	"p strictly implies (or: s-implies) q"
5. $[B] \mapsto$	doxastic implication	$p[B] \mapsto q$ iff $(p + B) \vdash q$	"p doxastically yields q, relative to B"

Note 1: As indicated above with enthymematic and doxastic "implication" ($[S] \mapsto$ and $[B] \mapsto$) the asserted propositions at issue can be seen as *putative* (rather than necessarily as *actual*) truths.

Note 2: We here suppose that the enthymematic set S and the doxastic set B are proportional sets that are closed under deduction.

Case 4   *S* is the set of logico-conceptually *necessary* truths (*N*), so that in general: $p \in S$ iff $\Box p$. Then $[S]\!\!\mapsto$ is strict implication $\prec$. So here $S = N = \{p: \Box p\}$.

Case 5   *S* is the set of our belief commitments (*B*). Then $[S]\!\!\mapsto$ is doxastic implication $[B]\!\!\mapsto$. (This issue has yet to be discussed.) So here $S = B = \{p: p$ is believed$\}$.[7]

Of course, the circumstance that a given statement actually follows from certain other particular statements is inevitably a matter of logic, a logical necessity. But that there are certain contingently characterized propositions (such as beliefs) from which a given fact follows is not. In consequence of this, most of these implication relationships ($\vdash$ and $\prec$ are excepted) will generally obtain only contingently.

It is also instructive in this context to consider the significance of universal impliedness:

$$(\forall p)(p \rightarrow q)$$

Note that we now have it that:

If $\rightarrow$ is $\vdash$, then $(\forall p)(p \vdash q)$ comes to $\vdash q$.

If $\rightarrow$ is $\supset$, then $(\forall p)(p \supset q)$ comes to q.

If $\rightarrow$ is $\prec$, then $(\forall p)(p \prec q)$ comes to $\Box q$.

If $\rightarrow$ is $[S]\!\!\mapsto$, then $(\forall p)(p\,[S]\!\!\mapsto q)$ comes to $q \in S$.

If $\rightarrow$ is $[B]\!\!\mapsto$, then $(\forall p)(p\,[B]\!\!\mapsto q)$ comes to $q \in B$.

Universal impliedness faithfully reflects the nature of the particular implication relationship that is at issue.

We can now state compactly the overall thesis on which the present analysis of conditionals is predicated, namely that while the basic idea of implication (symbolically, $p \rightarrow q$) encompasses a variety of constructions, nevertheless all of these trace back to one single fundamental idea, to wit, that of enthymematic implication. For we have it that:

$p \rightarrow q$ iff in the context of a suitably specified body of propositions S we have: $p\,[S]\!\!\mapsto q$, or equivalently, $(p + S) \vdash q$.

Thus in proceeding in terms of $[S]\!\!\mapsto$ we take a big step toward a unified theory of implication. For it is readily shown that all of the modes of implication listed in table 3.1 observe all these principles considered in the preceding section (implicativity, deductivity, reflexivity, etc.).

In their authoritative history of logic, the Kneales have written that "much confusion has been produced in logic by the attempt to identify conditional statements with expressions of entailment [i.e., with deducibility]."[8] But the present account will have it that such an identification is indeed possible provided that some due complications are introduced into those "expressions of entailment." The long and short of it is that implication as a generic idea admits of a substantial variety of implementations all of which can be synthesized under the unifying aegis of specifically *enthymematic* implication.

But let us examine somewhat more closely the specific differentiating features that characterize some of the key modes of implication.

6   Material Implication

Construing $p \to q$ as "Not-p or q" brings us to the special case of so-called *material implication* ($\supset$). Here we have:

$$p \supset q \text{ iff } \sim p \vee q$$

Clearly, if $\sim p \vee q$ is true, then q must be true if p obtains.

However, as an illustration of the oddity of material implication, consider the following scenario. You are in New York and ask your travel agent for a routing to Tokyo. He proceeds to sell you a ticket to Paris. You ask for an explanation, and he replies:

If you are in Paris, you are in Tokyo.

Your journey is a disappointment. Upon your return you charge him with deceit. He replies: "What I told you is the truth. We talked in New York, so that antecedent 'You are in Paris' was quite false. And, of course, a (material) conditional with a false antecedent is true." It is certain that neither you nor the judge or jury in your suit for fraudulent misrepresentation would be satisfied with that travel agent's explanation.

And so, since a true consequent is initially implied by anything, and moreover a false antecedent internally implies anything (*ex facto quod libet*), this mode of implication certainly does not enable us to capture the idea of conditionalization in general. Consider the conditional:

If Washington crossed the Delaware, then Lincoln wore a stovepipe hat.

This perfectly correct material implication would ordinarily be seen as absurd. For we standardly require more of plausible conditionals than mere truth-coordination. As P. F. Strawson puts it, we would ordinarily be prepared to say that a conditional is correct "only if we were also prepared to say that the fulfillment of the antecedent was, at least in part, the explanation of the fulfillment of the consequent."[9] This is certainly not the case with material implication. As one grammatical theorist writes: "*'If Paris is the capital of France, then two is an even number'* may well be acceptable from a logician's perspective [with "if-then" taken as material implication], but natural language conditionals generally demand a closer connection between *if*-clause and *then*-clause."[10]

Thus it is useful in this context to distinguish between two types of conditionals: correlation conditionals and connection conditionals. Consider the list: (1) Bob, (2) Tom, (3) Rob, (4) Ted. Here we have such conditionals as

If the name is even numbered, then it begins with "T."

If the name is odd numbered, then it ends with "ob."

There is—or may well be—no particular reason for this: it's just how matters happen to stand. Material conditionals are ideally suited to express the if-then at issue with such a correlation conditional. But the matter is quite different with:

If you are in Toronto, then you are in Canada.

Here there is a ground or reason behind the conditional, which pivots on the connection established through the fact that Toronto is in Canada. With such a substantively mediated conditional, an interpretation in terms of material implication fails to do justice to the matter.

The transitivity of implication may seem to be called into question by such examples as:[11]

If it is autumn, then that tree will shed its leaves.
If that tree sheds its leaves, then it is deciduous.
∴ If it is autumn, that tree is deciduous.

If I clap my hands, then that bird will fly off.
If that bird flies off, then it is not an ostrich.
∴ If I clap my hands, then that bird is not an ostrich.

The problem here, however, is not a failure of transitivity but rather a mismatched conditionality. Those premised implications are "natural" conditionals—that is, they have a causal or classificatory basis. But the only viable way to read the implication at issue in the conclusion is as a material implication ($\supset$). Now material implication is certainly transitive. But our inferences nevertheless seem anomalous because those "natural" implications in the premises lead us to expect more of the conclusion.

The crux is that material implication ($\supset$) is entirely truth-functional: the truth of $p \supset q$ hinges wholly on the truth status of p and q relative to the tabulation:

p	q	$p \supset q$
T	T	T
T	F	F
F	T	T
F	F	T

On the other hand, with any mode of implication stronger than $\supset$, $\rightarrow$ will be merely quasi-truth-functional, as per the tabulation:

p	q	$p \rightarrow q$
T	T	?
T	F	F
F	T	?
F	F	?

Here "?" stands for "true or false, depending on the circumstances," and so with any $\supset$-transcending implication we have it that when the antecedent is false all bets are off as far as the truth status of the conditional is concerned: be the consequent T or F, we can say nothing about the truth status of $p \rightarrow q$ without a deeper look at the specifics of the matter.[12]

The utility of material conditionalization lies in its applications in matters of deductive logic. As observed above, we have:

If $p \vdash q$, then $\vdash p \supset q$.

Moreover, the converse also obtains:

If $\vdash p \supset q$, then $p \vdash q$.

For suppose p. Then since (by the given antecedent) $p \supset q$ holds demonstrably, we can derive q. And thus $p \vdash q$ obtains: *demonstrable* natural implication thus deals correlatively with inferential derivability. And so as long as we confine our endorsement of material implication to *demonstrably true* instances, we can preserve a link between such implications and the deducibility that is quintessentially characteristic of authentic conditionalization.

Gottlob Frege's classic studies of the last quarter of the nineteenth century mark him as the father of material implication. He was followed in this regard by Bertrand Russell and the Ludwig Wittgenstein of the *Tractatus Logico-Philosophicus*, who argued vigorously for a purely truth-functional concept of implication. Certainly for the purposes of classical mathematics, it transpires that when $p \supset q$ can be established by way of proof this turns out to be a powerful instrument since, as we have just seen, it is tantamount to $p \vdash q$.

And so in mathematics and in deductive logic, where matters of demonstration and of derivability relationships are paramount, material implication suffices for all requisite purposes. (After all, *demonstrated* material implications are not merely that; they are also strict implications.) Since there are no demonstrable falsehoods, the principle *ex falso quodlibet*, which creates problems with contingent matters, has no chance to do damage here.

But this of course is not how matters stand in general. The theses

If q, then $p \to q$

If $\sim p$, then $p \to q$

have come to be called paradoxes of implication. They convey several instructive lessons, including the following two:

1. If $\to$ is construed as material implication, with $p \to q$ amounting to $\sim p \vee q$, then both "paradoxes" are inescapable and acceptable. However,

2. If $\to$ is construed as something stronger and more binding than $\supset$—as, for example, when it is construed as deducibility ($\vdash$)—then these paradoxes will not obtain. For example, to validate $p \to q$ for arbiting q we must have it that $\vdash q$ (that q is practicable) rather than simply that q, or "that q" is assertibly true.

It should also be noted in this connection that although the inference

$p \to (q \to r) \vdash (p \, \& \, q) \to r$

is clearly valid (since r becomes deducible from $p \, \& \, q$ when that antecedent premise is given), nevertheless its converse inference

$(p \, \& \, q) \to r \vdash p \to (q \to r)$

is not. Rather than being a general fact, it is an idiosyncratic characteristic of specifically *material* implication ($\supset$). Thus consider the conditional:

If r is a tree that sheds its leaves in the autumn, then x is deciduous.

Or, equivalently:

If x is a tree and x sheds its leaves whenever it is autumn, then x is deciduous.

This is equivalent to:

If x sheds its leaves whenever it is autumn and x is a tree, then x is deciduous.

This clearly has the format $(p \,\&\, q) \to r$. But consider its counterpart, of format $p \to (q \to r)$, namely:

(x sheds its leaves whenever it is autumn) $\to$ (x is a tree $\to$ x is deciduous).

This conditional is going to hold when $\to$ is material implication ($\supset$). But with any stronger implication it is going to come to grief over the fact that its consequent will then be false.

W. V. O. Quine has maintained that "whereas there is much to be said for the material conditional as a version of 'if-then,' there is nothing to be said for it as a version of 'implies.'"[13] But this is not quite correct. For whereas *demonstrably correct* material implications do indeed convey a peculiar sort of implicativeness, mere material implications as such will not even convey the idea of "if-then" as usually employed. In sum, "material implication" is a technical concept that has a life of its own, detached from any propositional relationships that have their natural home in ordinary language.

7 Demonstrability, Necessity, and Strict Implication

It is of interest to consider the connection between deducibility/demonstrability ($\vdash$) on the one hand and the modality of necessity ($\square$) on the other. Two considerations are basic here. The first is that deducibility and demonstrability are connected by the consideration that:

$\vdash p$ iff $\varnothing \vdash p$, where $\varnothing$ is the null set of premises or else a trivial premise on the order of $p \lor \sim p$.

And the second is that we thus do and must have:

If $\vdash p$, then $\square p$

And when conjoined these two yield:

If $\varnothing \vdash p$, then $\square p$.

On this basis, when a thesis can be proven on logico-conceptual principles alone, it is (demonstrably) necessary. This *provability principle* has it that demonstrability (provability) affords a prime pathway to necessity.

But what of the converse of this principle, namely

If $\square p$, then $\varnothing \vdash p$, or equivalently, $\vdash p$?

Effectively this states that the *only* pathway to necessity is via logico-conceptual demonstrability. And this is problematic, since it rules out other modes of necessity (say, physical, or natural, or stipulative), limiting necessity to logic-conceptual necessity alone.

The idea of inferential necessitation relative to a condition is conveyed by the locution "If p, then (relative to p) q obtains by necessity." This, of course, goes beyond if-then conditionality as such. In theory it can be construed in (at least) the five following ways:

(1) $p \vdash \square q$

(2) $p \vdash q$

(3) $\vdash (p \to q)$

(4) $\square(p \to q)$

(5) $\square p \vdash \square q$

In point of demonstrability we have the following chain of demonstrability:

$(1) \Rightarrow (2) \Leftrightarrow (3) \Rightarrow (4) \Rightarrow (5)$

For maximal plausibility in construing that initial locution, the extremes of (1) and (5) can be eliminated as too strong and too weak, respectively. The two other possibilities, namely (2)/(3), and (4), remain as potentially viable candidates for approximating the ordinary-language assertion "If p, then necessarily q," the latter construction being somewhat weaker than "If p, then (relative to p) q obtains by necessity."

Both (1) and (5) envision a necessity of consequent (*necessitas consequentis*) that renders them problematic. What is at issue with the other cases is a necessity of consequence (*necessitas consequentiae*) that meets the needs of the explanatory situation. (The medieval distinction at issue here is still able to do useful work.)

Here (4) provides an occasion also to consider *strict implication* ($\prec$) as instructed by C. I. Lewis. This is the implication relation that obtains when $p \supset q$ is a necessary thesis:

$p \prec q$ iff $\Box(p \supset q)$

Since we have

If $\vdash p$, then $\Box p$,

but not necessarily conversely, it thus transpires that $\prec$ is a relationship that is intermediate in strength and stringency between $\supset$ and $\vdash$.

8   Some Historical Background

The theory of conditionals was launched in ancient Greece by a trio of philosopher-logicians. In chronological order they were:

Diodorus Cronus (ca. 360–307 BCE), a pupil of Appolonius Cronus, himself a student of the famous paradoxer Eubulides.

Philo of Megara (ca. 330–270 BCE).

Chrysippus (ca. 280–205 BCE).

Although each one was prolific, none of their writings have survived. However, we know a great deal about them, principally from the writings of Diogenes Laertius, a historian of philosophy; from those of Sextus Empiricus, the skeptical Greek philosopher; and from those of Cicero, the politician and polymath.[14]

Two rival theories of conditionals evolved among the ancients which have come to be known as the Philonian and the Diodorean after their main exponents. The ancient discussion of the conditional was based on construing conditionals as if-then statements connecting complete propositions as per:

If Socrates comes, then Plato will converse with him.

Accordingly, to the later physician and sceptical polymath Sextus Empiricus, the disagreement between Philo and Diodorus stood as follows:

Philo says that a true conditional is one which does not have a true antecedent and a false consequent: for example, when both it is day and I am conversing, then "If it is day then I am conversing" is true. But Diodorus defines it as on which neither is nor ever was arguable of having a true antecedent and a false consequent. Accordingly to him, the conditional just mentioned would be false, since when it is day and I have become silent, it will have a true antecedent and a false consequent. But the following conditional would be true: "If atomic elements of things do not exist, then atomic elements of things do exist." For it will always have a false antecedent [so that it will never happen that the antecedent is true and the consequent false].[15]

Thus, for Philo, a conditional $p \Rightarrow q$ is true provided that $p \& \sim q$ is not *now* the case, while for Diodorus it is true only (but always) when $p \& \sim q$ is *never* the case. Both theorists seem to have had in view tensed (or temporalized) propositions of the type "Socrates is sitting [at time t]" or "It is day [at time t]." And so, for Philo:

$p \Rightarrow q$ iff For $t =$ now: Not both p-at-t and not-q-at-t,

whereas for Diodorus:

$p \Rightarrow q$ iff For every t: Not both p-at-t and not-q-at-t.

Both theorists thus agreed on an essentially truth-coordinated view of the conditionals as resting not on relationships but merely on coincidence times. The difference is whether the present situation or temporal generality across the board is at issue, Philonean implication being geared to the contemporaneous present while Diodorean implication is omnitemporal. In modern notation—with $R_t(p)$ representing "p is realized at time t"— Philonean implication comes to:

$\sim(R_n(p) \& \sim R_n(q))$,

whereas Diodorean implication comes to:

$(\forall t)\sim(R_t(p) \& \sim R_t(q))$.

The moderns, by contrast, have put the element of temporality aside. For them, the salient idea—that it is not the case that the antecedent is true and the consequent false—which was the common core of the Stoics' discussion, has been retained in material implication ($\supset$) subject to the specification:

$p \supset q$ iff $\sim(p \& \sim q)$, or equivalently $\sim p \vee q$.

And there is also the more stringent relationship obtaining when $\sim q$ is actually incompatible with p so that either $p \vdash \sim q$ or at least $\sim\Box(p \& \sim q)$.

Be this as it may, however, the quarrel between the Philonean and Diodorean theory of conditionals can be clarified on this basis. It pivots on the question of the validity of the inference:

$$\frac{\text{If } p \& q, \text{ then } r}{\therefore \text{ If } p, \text{ then if } q \text{ then } r} \qquad \frac{(p \& q) \Rightarrow r}{\therefore p \Rightarrow (q \Rightarrow r)}$$

On the Philonean construal of implication, this argument becomes

$$\frac{R_n(p \& q) \supset R_n(r)}{\therefore R_n(p) \supset [R_n(q) \supset R_n(r)]}$$

and this argument is clearly valid provided that we have: $R_n(p \& q)$ iff $R_n(p) \& R_n(q)$.

But on a Diodorean construction construal of implication we have:

$$\frac{(\forall t)(R_t(p \& q) \supset R_t(r))}{\therefore (\forall t)(R_t(p \& q) \supset R_t(\forall t')(R_t'(p) \supset R_t'(q)))}$$

That this is not going to work can be seen by the example of a three-period time-scale where the status of our propositions in point of obtaining (+) or not obtaining (−) stands as follows:

	p	q	r
(1)	+	−	+
(2)	−	+	−
(3)	+	+	+

Here the Philonean argumentation is just fine: no matter what time-period we pick for n—be it (1) or (2) or (3)—whenever the premise $(p \& q)$ is true, so is the conclusion (r). But the Chrysippean version of the inference fails owing to the falsity of the omnitemporal $(\forall t')(R_t^1(p) \supset R_t(q))$.

A yet different construal of conditionals is that of the following approach attributable to Chrysippus, the third member of the aforementioned trio.

Those who introduce connection or coherence [into the meaning of conditionality] say that a conditional holds whenever the denial of the consequent is incompatible with its antecedent.[16]

On such a view we have:

$p \Rightarrow q$ iff $(p \& \sim q) \vdash$ (contradiction).

On such a construction, Chrysippean implication $p \Rightarrow q$ is effectively tantamount to $\vdash (p \supset q)$.[17] And this, of course, envisions an entirely different approach to the matter. In effect, such a view of conditionals, unlike that of the Stoics, proceeds in terms of incompatibility and thus ultimately in terms of logical connection or derivability. So now truth-status coordination is no longer the pivot; rather, a far stronger conceptual connection comes to be required.

4   Conditional Complications

1   Rejecting Conditionals

Suppose that X says "If we offer him the bribe, he will take it," and Y, seeking to disagree, says "Not so: If we offer him the bribe, he will not take it." They deliberate and decide not to offer the bribe after all. That antecedent "We offer him the bribe" is accordingly false. But on the material implication construal of the conditionals, *both of those conditionals are true* (since then $p \supset q$ is true whenever p is false). This might seem to be a happy result for our disputants—they can have it both ways. But it hardly makes sense. On any common view of the matter they are in diametrical disagreement. So what is to be said about "conflicting" conditionals?

With universal statements of the form "All A is B" we must distinguish between the *contrary* ("All A is not-B" or equivalently "No A is B") and the *contradictory* ("Some A is not B") rejection. This distinction carries over to conditionals as well. For in the same way one must distinguish between the contradictory *negation* of a conditional, $\sim(p \Rightarrow q)$, on the one hand and on the other its contrary *denial*, $p \Rightarrow \sim q$. With conditionals, rejection is an adverbial process that can apply either consequentially or consequently. (Compare sect. 8 of chap. 2.) The result of drawing this distinction is depicted in table 4.1. It is clear that quite different things are at issue in each case.

Consider the situation obtaining when both $p \Rightarrow q$ and $\sim p \Rightarrow q$. Then q obtains irrespective of the status of p, so that p is simply irrelevant to q. In this case q itself could and should be asserted directly; there is nothing conditional about it. On this basis there is not much point to the conditional contention:

If today is Tuesday, then $2 + 2 = 4$.

Table 4.1
The Negation and Denial of Different Implications

Implication relation	Negation	Denial
$p \Rightarrow q$	$\sim(p \Rightarrow q)$	$p \Rightarrow \sim q$
$p \vdash q$ (deducibility)	not: $p \vdash q$	$p \vdash \sim q$
$p \supset q$ (material implication)	$p \;\& \sim q$	$p \supset \sim q$ (or $\sim p \vee \sim q$)
$p \prec q$ (strict implication)	$\Diamond(p \;\& \sim q)$	$\sim\Diamond(p \;\& q)$
$p\,[S]\!\mapsto q$ (enthymematic implication)	Not: $(p + S) \vdash q$	$(p + S) \vdash \sim q$

For $2 + 2 = 4$ obtains irrespective of the antecedent, and indeed we could rightly say something on the order of "Be today Tuesday or not, still [or: *nevertheless*] $2 + 2 = 4$." However, in such a situation the conditional at issue is to be dismissed as badly framed or even pointless rather than false or erroneous.

2   Complex Implication Conditionals

One contemporary logician says that "we rarely have in natural language any use for conditionals whose antecedents are themselves conditionals."[1] But this is questionable. That conditioned conditionals are nothing all that unusual is shown by such examples as:

If people are mistaken when they believe him, then I too am in error.

If those are sinners who tell less than the whole truth when asked, then most of us are.

The fact is that ordinary-language statements frequently involve conditionals within conditionals. For example, "He can do it, if she helps him" comes to something like "If he endeavors to do it, then if she helps him, he will be able to manage it." Such statements in effect contain if-then statements in their antecedents or consequents.

Again, consider a disposition-characterizing statements such as:

If something is made of sugar, then it is water-soluble.

This comes to something like:

If an item is made of sugar, then: if this item is placed in water, then it will dissolve.

There is no shortage of conditioned conditionals of this general kind.

3 A Digression on Monotonicity and Transitivity

To qualify as *monotonic* the "implication" at issue in a conditional must be airtight: it must authorize us to stake a claim on the order of "If p, then for sure q." Accordingly, the monotonicity-characterizing principle is:

If $p \Rightarrow q$, then $(p \& r) \Rightarrow q$ for any and every r.

No qualification additional to the antecedent as such can abrogate a valid monotonic implication: the antecedent will, in and of itself, suffice to guarantee the consequent. Whenever "inevitably (invariably, unavoidably, etc.)" can be weakened to "generally, usually, probably, possibly, etc.)," the monotonicity that is requisite for authentic implication is lost.

However, consider:

If you are in America, then you might be in New York.

This is, of course, perfectly correct. But it will not do to "strengthen" the antecedent as per:

If you are in America and you are in Texas, then you might be in New York.

Since monotonicity fails us here, the conditional at issue here will be a mere pseudo-implication, seeing that authentic implications must be monotonic. For example, suppose that a certain sort of conditional if-then is nonmonotonic. Then for some p, q, r we have

$p \Rightarrow q$ but not $(p \& r) \Rightarrow q$

Now if $(p \& r) \Rightarrow q$ fails, then $\sim q$ must be compatible with $p \& r$. But when the implication is contraposable, then $p \Rightarrow q$ yields $\sim q \Rightarrow \sim p$. Hence $\sim p$ should be compatible with $p \& r$, which it is not. So nonmonotonicity is incompatible with contrapositivity, and contraposable conditionals must be implicationally monotonic.

The fatal defect of such nonmonotonic pseudo-implications from the standpoint of deductive logic is that we lose the crucial principle of *modus ponens*: when p and $p \Rightarrow q$ are given we cannot securely infer q. For we lack the assurance that there is not in the background some true r which is such that q can fail when $p \& r$ obtains.

Ordinarily we would not hesitate to argue from descriptive generalizations as per "As are Bs" to conditionals as per "For any x, if x is an

A, then *x* is a *B*." But problems can arise here. Although we do indeed have

Birds are flyers (i.e., creatures that can fly)

and

Owls are hooters (i.e., creatures that can hoot)

we do not have

Wingless birds can fly

or

Stuffed owls can hoot.

The problem, however, is that a loosely formulated everyday generalization like "Birds can fly" does not actually encode "All birds whatsoever can fly" but merely "*All typical* birds can fly." And, of course, when Ozzie is an ostrich the reasoning

All typical birds can fly.
Ozzie is a typical bird.
∴ Ozzie can fly.

collapses owing to the failure of its second premise. (And a strictly analogous course of reasoning removes the difficulty in the case of Ollie the Owl.)

Accordingly, what the preceding examples reveal is not so much a failure of monotonicity for generalization-based implications as a failure of strictly correct formulation. For many generalizations are governed by a presupposition of normalcy, and this is simply not forthcoming in such cases.[2]

A parallel line of thought holds with regard to transitivity as well. Thus consider the Job's comforter who presents us with the idea that we don't have to worry about losing our purse thanks to the following argument:

If you lose your purse, then you are penniless.
If you are penniless, then there is no money in your purse.
∴ Not to worry if you loose your purse, for if you do so then there's no money in it.

Analogous instances of a "failure of transitivity" arise in examples like:

If Germany had won World War I, then World War II would not have occurred.
If World War II had not occurred, Hitler would have ruled the continent.
∴ If Germany had won World War I, Hitler would have ruled the continent.

If I were president of General Motors, I would be a rich man.
If I were a rich man, I would drive a Rolls-Royce.
∴ If I were president of General Motors, I would drive a Rolls-Royce.

All of these inferences go amiss over the same fact, namely that the requirement of normalcy or "other things equal" on which they hinge goes wrong. In each case, the two different conditionals at issue have discordant enthymematic bases.

4 Quantificational Complications

Typically, if-then conditionals connect completed statements ("If today is Sunday, then tomorrow is Monday").[3] But often this is not the case, and there is a quantifier-demanding cross-reference from one component of the conditional to the other, as per:

If a tree is deciduous, then it will shed its leaves in autumn.

The logical structure of such a statement is:

$(\forall x)(Tx \Rightarrow Sx)$

Or again, consider:

If someone hesitates to respond to a question, then he is uncertain of the answer.

If the buyer is a member, then she will get a discount.

If a bird is an owl, then if it is making the sounds that are natural to it, then it will hoot.

The logical form of these three implication statements is not $p \Rightarrow q$, but, respectively:

$(\forall x)(\forall q)(Hxq \Rightarrow Uxq)$

$(\forall x)((Bx \,\&\, Mx) \Rightarrow Dx)$

$(\forall x)(Ox \Rightarrow (Nx \Rightarrow Hx))$

Moreover, there is no requirement for these cross-referential statements to be universal. Thus, consider:

Some staff member will leak the memo if asked. Symbolically:
$(\exists x)(Sx \,\&\, (Ax \Rightarrow Lx))$

Neither the antecedent nor the consequent of such conditionals is a stand-alone completed proposition because those quantifiers reach across and connect the statement's components. With such cases of "quantifying in" there is no question of a $p \Rightarrow q$ format. Conditional if-then statements of this general sort where propositional functions rather than complete statements are at issue may be called *quasi-conditionals.*

But how is the deductive-enthymematic analysis of substantive conditionals, geared as it is to the format $p \Rightarrow q$, to be applied in such cases? The answer is that a modest bit of supplementary sophistication is needed.

To get the deductive pattern to apply in such cases we need to introduce the idea of a parametric pseudo-constant which can be either universal $u, u_1, u_2, u_3 \ldots$, or particular $s, s_1, s_2, s_3 \ldots$. These pseudo-constants can be used to encode our information at the start of the deductive process and subsequently on its eventual completion they can be decoded appropriately. For example, consider the quasi-conditional:

If a member of the club requests it, they will be admitted. Symbolically:
$(\forall x)((Mx \,\&\, Rx)\,[S]\!\!\rightarrow Ax)$.

Using the indicated symbolic convention we recast this as:

$(Mu \,\&\, Ru)\,[S]\!\!\rightarrow Au$

And we now proceed in the usual way noting that:

$((Mu \,\&\, Ru) + (\text{Background information as to the linkage of admission to club membership})) \vdash Au$

Or again, consider:

If asked to leak the memo, some staff member or other will do so. Symbolically: $(\exists x)(Mx \,\&\, (Ax\,[S]\!\!\rightarrow Lx))$

As per convention we will here posit Ms and further take recourse to our background information to establish:

(*As* + Certain background information about the indiscretion of staff members) ⊢ *Ls*

This validates:

Ms & *As* [*S*]↦ *Ls*

And now we generalize existentially as per our convention to obtain the desired conclusion:

(∃*x*)(*Ms* & *Ax* [*S*]↦ *Lx*)

In sum, by introducing the convention of quantificationally geared pseudo-constants in place of quantified variables it becomes possible to keep the analysis of such quantified quasi-conditionals within the framework of our standard enthymematic-deductive account.

5 Problems of Probability

In view of the distinction between consequent and consequential qualifications, it is readily seen that there is a significant difference between

$p \Rightarrow$ likely q

and

likely ($p \Rightarrow q$).

These are certainly not equivalent claims. Thus let p be true but unlikely, while q is also unlikely. Then "$p \Rightarrow$ likely q" must be false (since its antecedent is true and its consequent false). Nevertheless, $p \Rightarrow q$ can readily be likely, since there is no problem about one unlikely proposition entailing another. (In particular, note the situation that arises when $p = q$!)

Ernest W. Adams has suggested that "the probability of an indicative conditional of the form 'if A is the case, then B is' is a conditional probability."[4] On this approach we have:

$pr(p \Rightarrow q) = pr(q/p)$

But this idea encounters problems.

To begin with let $\Rightarrow$ be material implication ($\supset$). Then

$pr(p \supset q) = pr(\sim p \vee q)$

Now let us toss a normal die, and let

$p = 1$-or-2 comes up

$q = 1$ comes up

Then $pr(q/p)$ is $\frac{1}{2}$ while $pr(p \supset q)$ is $\frac{5}{6}$.

Again, if $\Rightarrow$ is deducibility ($\vdash$), then since q is flat-out nondeducible from p, $pr(p \vdash q)$ is 0, which is not the case with $pr(q/p)$.

All in all, the probability of conditionals lacks a stable relationship to the conditional probability involved.

One might consider the idea of a conditional defined in terms of high conditional probability, along the lines of:

$p \gg q$ iff $pr(q/p) > .9$

However, such probabilistic conditionalization will certainly not represent a mode of implication. Among other things, it is subject to neither *modus ponens* nor transitivity. (Chapter 12 below will return to some of the issues that are relevant here.)

5 Doxastic Implication and Plausibility

1 Substantive Implication and Its Tenability

Doxastic implication is a special version of enthymematically substantive implication, namely one relativized to accepted beliefs. Thus letting B be the manifold of one's issue-relevant *belief commitments*[1]—which we shall suppose to be collectively consistent—we arrive at the specification:

$$p \; [B] \!\mapsto q \; \text{iff} \; (p + B) \vdash q$$

Thus when $p \; [B] \!\mapsto q$ obtains, then someone with the belief manifold B will be (logically) committed to q in also accepting (or supposing) that p.

For example, consider the conditional:

If they invite me to the party, I shall accept.

The background information of relevant beliefs that provides the enthymematic basis of this conditional presumably stands somewhat as follows:

I want to go to the party and stand ready to do what is required for this.

Going to the party requires getting an invitation.

Now suppose that they invite me, thereby offering me a chance to go. Since—by hypothesis—I am set to avail myself of such a chance, my acceptance of the invitation follows. In constituting the enthymematic basis of the conditional, belief commitments provide for a deductive linkage between the hypothetical antecedent and the indicated conclusion. In this way, the tenability of a doxastic conditional pivots on the conjunction of the relevant background beliefs.

Since this conception of doxastic implication proceeds with a view to a background of belief—that is, with reference to *putative* truth rather than true as such or *actually correct* beliefs—it represents an epistemic rather than purely logical relationship.

Doxastic implication represents what has becomes known as a *supposi-tional* construal of implication. This approach goes back to a 1929 paper by F. P. Ramsey, who maintained:

> If two people are arguing about "If p, then will q?" and are both in doubt as to p, they are adding p hypothetically to the stock of their knowledge, and arguing on that basis about q; ... they are fixing their ... belief in q given p.[2]

Such a linkage to accepted beliefs means that doxastic implication is a re-source that belongs to the realm of information management rather than to abstract logic as such. It represents what is, in the final analysis, not a strictly semantic concept but a resource of applied rather than theoretical logic. All the same, doxastic implication is closely geared to such a logic, and in particular it is clear that $p \vdash q$ will yield $p\,[B] \!\!\mapsto q$ irrespective of the composition of B.[3]

Some points regarding the relationships among different types of be-lievers deserve note. Consider the thesis:

$$p\,[B] \!\!\mapsto q \text{ iff } p \supset q$$

This would hold in general for any believer who believes *all* true proposi-tions (i.e., is omniscient) so that $p \in B$ iff p. But it holds only for such believers, and certainly not in general.

On the other hand, consider the thesis:

If $p\,[B] \!\!\mapsto q$ then $p \vdash q$

This would hold in general for any believer who is totally agnostic apart from logico-conceptual matters, that is, believes *nothing* outside the range of logico-conceptual truth, so that $p \in B$ iff $\vdash B$.

Does doxastic *implication* actually merit this designation: is it really a form of implication? Thus suppose that we believe $p \supset q$ so that $(p \supset q) \in B$. Then clearly $p\,\{B\} \!\!\mapsto q$. But any generic mode of implication must obey *modus ponens*. And yet we here do not really have an inference from p and $p\,\{B\} \!\!\mapsto q$ to q, but only to $q \in B$ for logically competent believers. We are, in sum, dealing not with the truth per se, but with *pu-tative* truth: with what is (or should be) believed to be true by the doxastic subject at issue.

The sensible way to handle this observation is to acknowledge that the situation is as it seems, that what is at issue here is not the absolute truth, but the truth as we see it, the *putative* truth, the only sort of truth with which we can conceivably come to grips. And then we can and should also move on to say that if *modus ponens* as traditionally construed is to count as a requisite for implication proper, the doxastic "implication" can only be counted as a quasi-implication of sorts.

2 What Can Go Wrong with Doxastic Implications?

There is a radical asymmetry between validating a doxastic conditional and invalidating it—that is between establishing its truth and its falsity. For to establish

$$p \, [B] \!\!\mapsto q$$

or equivalently

$$(p + B) \vdash q,$$

it suffices simply to adduce some B-member (or conjunction thereof) such that $(p + [\text{this material}]) \vdash q$. However, *negating* such a conditional is a more complicated matter given that it involves a nonexistence claim. In rejecting doxastic conditionals we are therefore more likely to resort to denial than negation.

Against this background, it is germane to ask: What can be said regarding the pathology of doxastic implication? What sorts of failure can block the way to the tenability of $[B] \!\!\mapsto$-style conditionals? How can they malfunction?

Consider once more the defining idea of a doxastic implication:

$$p \, [B] \!\!\mapsto q \, \text{iff} \, (p + B) \vdash q$$

It is clear that two sorts of things can go wrong here:

1. A defect of information—that is, a B-defect. The information required to provide for q's derivation is not actually B-available: we are wrong about its presence in this belief manifold. Here, then, there is a deficit of B-afforded information.

2. A defect of inference—that is, a $\vdash$-defect. We are mistaken in thinking that the conjunction of our antecedent with B-information does actually entail the consequent. Here, then, there is an error in point of derivability.

A flaw of the first kind—that is, of information—occurs in such cases as when we mistakenly think that Dubuque is in Connecticut and then proceed to say:

Since Dubuque is in Connecticut, it is in New England (seeing that Connecticut is in New England).

The enthymematic basis of this conditional—namely, "Dubuque is in Connecticut"—ought not to be B-available.

A flaw of the second kind, one of inference, occurs in such cases as when we say:

If he was born in 1930 he must now (2006) be a septuagenarian (seeing that 2006 minus 1930 is 76).

The arithmetical inference purportedly at work here is simply not valid.

Doxastic conditionals can also go awry in yet more bizarre ways. Thus consider:

If the antecedent of this conditional is true, then its consequent is false.

If a conditional is to be tenable, then when its antecedent is true, its consequent must be true as well. Accordingly, there is no way for the contemplated conditional to be true, seeing that what it states saws the limb off on which its tenability hangs. What we have here is an outright paradox. To assert a statement (be it conditional or not) is to claim its truth, and such a claim just cannot be warranted here. In this (paradoxical) regard, the self-referential conditional at issue resembles the self-defeating contention: "This statement is false."

Some further complications must be noted with respect to doxastic conditionals. They pivot on the question of what it is that blocks the route to such substantively silly conditionals as:

If Paris is in France, then sugar is sweet.

After all, there is a true (and acceptable) proposition, namely

⟨Sugar is sweet⟩-or-⟨Paris is not in France⟩

from which, in conjunction with its antecedent, our conditional's consequence follows.

The problem here lies in the consequent's following from (antecedent $+ B$), because it follows from B alone—the antecedent has nothing to do with it. If doxastic conditionals are to convey serviceable information, complications of this sort must also be sidelined.

Again, consider the situation where we accept p, and *a fortiori* also $p \vee q$ for arbitrary q. Then we will have:

$$\sim p\left[\{p, p \vee q\}\right] \mapsto q$$

In this way, we would obtain *the anomalous*

If Paris is in Germany, then salt is sweet.

The problem here is that the antecedent ($\sim p$) is incompatible with the enthymematic background at issue (B). This case too must be sidelined. (It poses the problem of counterfactuals, which will be dealt with below.)

3 The Plausibility of Beliefs

With doxastic conditionals we enter a realm of substantive information rather than abstract logic alone. The derivation at issue with $(p + B) \vdash q$ is bound to hinge on the constitution of the manifold of p-relevant belief (B). And this circumstance endows doxastic conditionals with a characteristically epistemological feature that deserves further scrutiny—the issue of plausibility.

The plausibility of a belief is a reflection of its epistemic priority status—a matter not of its explicit substantive content but of its epistemic standing in the cognitive scheme of things. It indicates how central a role a contention plays within the wider framework of our cognition. The pivot here is the strength of our epistemic commitment to that belief.

A thesis is minimally plausible when it is merely *compatible* with what we accept—when it is true for aught we know. Beyond that, the plausibility of a thesis hinges on the systemic standing that we take a claim to occupy in the cognitive domain at hand; it is not a matter not of assessed probability or likelihood, but of standing or status in the systemic setting of our knowledge. (Think here of the analogy of an axiom in mathematics or formal logic; axiomaticity is not a matter of what the proposition *says* but rather of its *cognitive role* or *status* in the relevant system of propositions.)

In this regard, the order of decreasing priority stands as follows:

(a)   Definitions and acknowledged conceptually necessary theses (linguistic conventions, mathematical relationships, and principles of logic included).[4]

(b)   Basic ground rules and principles of rational procedure in matters of inquiry and world outlook.

(c)   Firmly entrenched observational or experiential "facts of life" regarding the world's ways: inductively grounded laws and well-confirmed lawful generalizations.

(d)   Well-substantiated commitments regarding particular matters of contingent fact.

(e)   Reasonably warranted contentions (first) about general relationships and (second) about specific facts.

(f)   Merely provisional assumptions and tentative conjectures about general relationships and (secondarily) about particular matters of fact.

(g)   Speculative suppositions about general relationships and (secondarily) about particular matters of fact.

The middle range of this register (c–e) is occupied by the sorts of propositions that Aristotle called *endoxa* in the opening chapter of his *Topics*—that is to say, generally acknowledged facts and widely accepted convictions. (This consideration rightly indicates once more that it need not be specifically *truths* that are at issue; here *putative* truths are once more at issue.)

Such a plausibility ordering reflects certain general principles of precedence through prioritizing

• the comparatively more basic and fundamental;
• the comparatively more general and far-reaching;
• the comparatively more factual and less conjectural;
• the comparatively better-evidenced and more reliable; and
• the comparatively more context-consonant and less far-fetched.

To reemphasize: what matters for plausibility is the systemic enmeshment or fundamentality of our beliefs within the wider setting of our cognitive commitments. It is a matter of the extent to which loss of the item in question would cause rends in the overall fabric of our beliefs.

However, while plausibility reflects cognitive commitment, this should be conceived of as not a matter of measurable degree (like probability) but one of stratified levels arrayed somewhat as follows:

I.   Accepted unqualifiedly as flat-out true—items (a)–(c)

II.   Highly plausible—item (d)

III.   Moderately plausible—item (e)

IV.   Somewhat plausible—item (f)

V.   Minimally plausible—item (g)

For all (or most) practical purposes we can get by with such a grading of plausibilities in terms of these five numerally indicated levels. And, of course, what might perhaps be thought of as the bottom end of such a scale—propositions deemed highly implausible and flat-out rejected as false—should be omitted here since propositions that are so considered would hardly be qualified to count among our beliefs.

Construed on this basis, precedence and priority in point of systemic plausibility is determined by the criteria and principles inherent in the teleological structure of factual inquiry, namely to aid us in doing the best we can to secure viable answers to our questions about how matters actually stand. Plausibility is, in effect, an index of cognitive status. The more plausible a thesis is, the greater the price of its abandonment would be—that is, the larger the extent to which its loss would have negative repercussions in the cognitive scheme of things as we see it. In forgoing a plausible thesis we always lose something we would (ideally) like to have. Plausibility is, to reemphasize, an index not of likelihood but of cognitive centrality or standing.

Accordingly it is important to realize that plausibility differs from probability in various ways but above all in point of conjunctivity. *For the plausibility status of a conjunction of accepted beliefs is fixed as the plausibility of its least plausible premise.* This is the crucial *weakest link principle* of plausibility determination. It means that a conjunction of beliefs that have same-status plausibilities will always be of this same status—decidedly unlike the situation with same-status probabilities.[5]

4   The Plausibility of Doxastic Conditionals: The Weakest Link Principle

The concept of plausibility is readily extended from categorical propositions to substantive conditionals. For the plausibility of a substantive conditional is readily determinable in terms of the plausibility of the enthymematic premises that support them. Specifically: *the plausibility of a substantive conditional is that of the weakest premise in the strongest argument for its validation via deduction from beliefs.* Accordingly, the plausibility of a substantive conditional

$$p\,[B]\!\!\rightarrow q, \text{ or equivalently, } (p + B) \vdash q$$

is determined by the plausibility of the least plausible B-member that is enthymematically required for the deduction of q from p. This, then, is the crux of the weakest link principle for analyzing the cogency of conditionals.

For the sake of illustration, consider the conditional:

If you drive faster than the speed limit, then you will get a ticket.

Here we have:

Beliefs

p = You drive faster than the speed limit. (Hypothetical supposition)

b_1 = Unlucky people who drive faster than the speed limit get ticketed. (Belief of level II plausibility)

b_2 = You are an unlucky person (in such matters as being ticketed). (Belief of level III plausibility)

q = You will get ticketed.

The tenability of this substantive conditional roots in the consideration that $(p \, \& \, (b_1 \, \& \, b_2)) \vdash q$ obtains under these circumstances. But note that while b_1 is highly plausible (relative to one's understanding of "unlucky"), b_2 is only somewhat (moderately) plausible. On this basis, the conditional as a whole would be moderately plausible, seeing that b_2, a level III thesis, represents its weakest link.

Of course there are, in theory, other ways of connecting the antecedent and consequent of an implication, in this case—as in all others. For example, one might contemplate recourse to:

b_3 = Everyone who drives faster than the speed limit gets ticketed.

But this is decidedly implausible, and the appropriate plausibility of a conditional is that of the most favorable case. In constructing enthymemes what we are after is maximum plausibility. And here the crux is a matter of *maximizing the minimum plausibility of the available connecting belief-premises*. We arrive at the general rule:

The plausibility of a substantive conditional is not determined by the plausibility of the antecedent and that of the consequent viewed in isolation. Instead, it depends entirely on the plausibility status of the least plausible among the belief-available enthymematic premises able to close the inferential gap between antecedent and consequent.

On this basis the tenability of a doxastic conditional will hinge on the plausibility status of the enthymematically operative beliefs so that the acceptability status of such conditionals is grounded in epistemic rather than merely semantical considerations.

6 Inferentially Insuperable Boundaries and Homogeneous Conditionals

1 The Conceptual Background

A deductive conditional as per $p \vdash q$ establishes an especially close informative bond between its component propositions. And in this light, the prospect of deductive relationships reaching across the boundaries between different propositional domains poses some interesting issues of theoretical analysis. In particular, it raises the question of whether deductively valid conditionalizations are *homogeneous* in confining the consequent to the same taxonomic region as the antecedent. This question is the topic of the present chapter.

The purely truth-functional nature of material implication means that the conditional $p \supset q$ can obtain even when p and q have nothing whatever to do with one another. Thus we have as a material conditional:

If Caesar crossed the Rubicon, then Napoleon died on St. Helena.

Strange, but true! However, matters stand very differently with the implication relationship at issue in logico-conceptually deductive entailment ($\vdash$). Here a far closer relationship clearly is needed. But must there be a type-kinship of thematic homogeneity between antecedent and consequent when this particularly strong sort of relationship obtains?

It would seem that the answer is a resounding affirmative. After all, deductive processes are informatively transformative rather than creative: whatever information emerges in the conclusion must somehow be provided by the premises. Accordingly, *deduction* must somehow connect like with like. We cannot deduce geographical truths from (merely) geometrical premises, nor chemical truths from (merely) arithmetical premises. It would seem that deduction cannot make inferential leaps across thematic boundaries. But the reality of the matter is more complicated than that, for the exact nature of the property-type taxonomy at issue has come into it.

A propositional domain D is inferentially closed with respect to a feature F whenever it transpires that F-characterized propositions in D can deductively yield only F-characterized conclusions. Such a domain is thus characterized by the principle:

For any p and q in D, and any r for which not $\vdash r$, if $F(p)$ and $F(q)$, and $(p \,\&\, q) \vdash r$, then $F(r)$.[1]

This, of course, also means (via the substitutions p/q and q/r) that

If $F(p)$ and both: not $\vdash q$ and $p \vdash q$, then $F(q)$.

What we have here is in effect an inferential barrier that impedes any deductive transit from F to (nontrivial) non-F. Put figuratively, we might say that F-characterized propositions will breed homogeneously true to type in deductive processes: when F-characterized premises go in, then (trivialities apart) F-characterized conclusions must result.

For the sake of illustration, note that a logically valid process of deductive inference will never permit crossing the barrier from fact to error across the true–false divide. No valid deduction ever leads from truths to falsehoods. However, the reverse situation is very different. A deductive transit from falsehood to truth is readily possible. (For example, the statement "George Washington was a Swedish general" is false, but it entails "Washington was a general," which is, of course, true.)

Let us consider some other dichotomies. The idea that a deductive barrier is set by the general–specific or universal–particular distinction is a point of contention between classical and modern logic. Classical logicians saw the inference from the affirmative universal "All S is P" to the particular "Some S is P" as valid. But modern logicians, who do not credit universal statements with existential import, see such an inference from generality to specificity as invalid, save under rather special conditions. (It is viewed as being, at best, enthymematically valid, with "There are Ss" as a requisite auxiliary premise in the present case.) On the other hand, all parties would agree on barring the reverse transit, regarding a deduction from mere particularities to (nontrivial) universalities as invalid.

Again, consider the boundary between the actual and the merely possible. Clearly, the logical consequence of any proposition that states a true fact must itself be one. Of course, whatever is is *ipso facto* possible, but there is no deductive transit from the actual to the *merely* possible: valid deductions with only actualistic premises will *ipso facto* have to yield actualistic conclusions.[2] Again, medieval logicians viewed the boundary set by the necessary–contingent distinction as deductively insuperable. John Duns Scotus, for example, maintained that no contingent truth ever follows from necessary premises alone.[3] This stance was not entirely uncontested, since orthodox doctrine saw God's existence and his activity as benevolent creator as necessary facets of his nature, while the world's existence—which seemingly follows from this—should be seen as contingent. However, modern logicians agree with Scotus that the logical consequence of any necessary proposition is itself bound to be necessary. But once again, the matter stands differently in the reverse direction; the consequence of a contingent proposition can be necessary, given that although p may be contingent, its logical consequence p-or-not-p will, of course, be necessary.

What about the distinction between "is" and "ought"—between factual and normative? Many philosophers take this to constitute a deductively insuperable barrier. And yet one wonders. The inferential step from "ought" to "is" looks to be quite simple: "People ought to recognize the truth of *Paris is the capital of France*" is a normative statement that carries the factual "Paris is the capital of France" in its deductive wake. However, the converse inference to norms from clear-cut facts is more problematic. Thus even though the fact that a certain answer to a certain question is false seems to entail that people ought to avoid accepting it as correct, still this follows only in the presence of the further normative truism that one should not deceive people (be it oneself or others).

Or consider the fact–value distinction. Many philosophers hold that facts cannot be inferred from values (or the reverse). Nevertheless, the evaluative contention that "It is unfortunate that p is true" surely entails "p is true." And "It is a good thing that cats chase mice" looks to be evaluative and yet entails the factual "Cats chase mice."

Nor does a deductive transit from factual to evaluative propositions seem infeasible. If f is a statement whose status is factual, then so surely is not-f. And yet if v is some arbitrary value thesis, then the factual thesis f will entail f-or-v which together with the factual claim not-f will entail v, which is, by hypothesis, evaluative. More concretely, the factual "All boys like (all) candy" entails both "All good boys like candy" and "All boys like good candy," which certainly look to be evaluative.

The barrier from subjective to objective seems harder to cross. One cannot easily get from what is liked by people to what is likable as such, or from what they accept to what is genuinely acceptable. In fact the leap across this sort of barrier has become known among philosophers as Mill's Fallacy.[4]

The analytic–synthetic distinction took some hard knocks in twentieth-century philosophy. Nevertheless there does seem to be a difference between those statements whose verification requires one to look at how things work in the real world ("All dogs have livers") and those whose verification requires only an inquiry into the usage of words ("All forks have tines"). So let us for the moment accept this distinction as cogent. Then, of course, the deductive transit from synthetic truths to analytic ones is easy: However synthetic p may be, its deductive consequence p-or-not-p had best be classed as an analytic truth. However, a deductive transit in the opposite direction would indeed seem more difficult. Whatever follows logically from the analytic propositions inherent in the conceptual fabric of language is itself bound to be analytic in the presently relevant sense.

Again, consider the temporal–atemporal-or-omnitemporal distinction that is at work in the contrast between "A copper sphere is present in this room" and "Copper conducts electricity," where the former obtains right now and the latter obtains always. Omnitemporal premises can obviously yield temporally qualified conclusions since whatever is so always will be so now. On the other hand, as long as a necessary proposition is derivable from any premise (i.e., as long as we have: If $\vdash p$, then $q \vdash p$, for any q) it will also be the case that temporal premises can yield atemporal consequences (such as $2 + 2 = 4$), though this becomes problematic when logico-conceptual necessities are excluded.

Again, the domain of the predictable is inferentially closed—the logical consequences of predictable facts are certainly also predictable. However, the reverse is not the case: unpredictable circumstances can have predictable consequences. For even where one cannot predict the year of a newborn child's death one can safely predict that it will be dead two hundred years hence.

Then, too, in the context of philosophical deliberations it could plausibly be argued that an inferential barrier separates mental and physical discourse. And in consequence it could be contended that there just is no conceptually secure transit from talk about thoughts and feelings to talk about physico-chemical brain processes.[5]

On this basis one can see that certain domains are deductively closed, in that the boundary between domain-internality and domain-externality is inferentially insuperable.

It is thus clear that various distinctions give rise to inferentially insuperable barriers—sometimes unidirectionally and sometimes bilaterally (as per universal–particular). And it is equally clear that whenever such insuperability obtains with regard to a distinction, we have a circumstance whose role in conceptual phenomenology is bound to be instructive.

2 The Logical Lay of the Land

Returning to the idea of inferential closure, let us say that a propositional feature F is *inferentially transmissible* over a propositional domain D when this domain is inferentially closed with respect to F, that is, when for all D members:

For any p and q, and any r such that not $\vdash r$: If $F(p)$ and $F(q)$ and $(p \,\&\, q) \vdash r$, then $F(r)$.

This, of course, also means (via the substitutions p/q and q/r) that

If $F(p)$ and not $\vdash q$ and $p \vdash q$, then $F(q)$.

A feature with this characteristic may be designated as *deductive*. With respect to such features, deductive inference is taxonomically homogeneous.

Deductivity is clearly going to hold for many propositional modifiers, for example: "is true," "is necessary," or "is possible." But it is clearly going to fail for others such as the modal "is *merely* possible," the semantic "is false," or the evaluative "is to be welcomed."

A feature F is *conjunctive* over a proptional domain D when for all D members:

If $F(p)$ and $F(q)$, then $F(p \,\&\, q)$.

Such conjunctivity clearly holds for any feature that is inferentially transmissible. (This follows by letting $r = (p \,\&\, q)$ in the specification above.)

Now, when the converse relationship holds, we have

If $F(p \,\&\, q)$, then $F(p)$ and $F(q)$.

In this event, the feature at issue is said to be *distributive*. Clearly any feature that is deductive—and *a fortiori* any that is inferentially transmissible—will also be distributive.

Accordingly, the preceding interrelationships indicate that:

(deductive + conjunctive) $\Leftrightarrow$ inferentially transmissible

deductive $\Rightarrow$ distributive

It is clear on this basis that certain logical interrelationships have to hold among the features with which we have been concerned.

7 Counterfactual Conditionals and Their Problematic Nature

1 Counterfactuals

Let us now enter into the region of cognitive pretense and make-believe— of "what-if." After all, things might have been very different than they are. Caesar might not have crossed the Rubicon. Napoleon might never have left Elba. Surely we can reason sensibly from such straightforward contrary-to-fact assumptions so as to obtain instructive information about the consequences and the ramifications of these unrealized possibilities. There can be little question that we can generally say something about "what would happen if" in such matters. We know perfectly well that (for example): "If Hitlerite Germany had developed an atomic bomb by 1943, World War II would have taken a very different course." What we have here is a *counterfactual conditional*: an answer to a question that asks what would transpire, were this antecedent is false—or judged to be so.[1] Just as explanations answer "why?" questions, these conditionals answer "what if?" questions. Accordingly, there are both falsifying and "truthifying" counterfactuals, respectively coordinated with the questions:

If p (which is true) were false, then what?

If p (which is false) were true, then what?

Answering such falsifying and truthifying questions is the very reason for the being of counterfactual conditionals.

A conditional is *speculative* when we view its antecedent in a neutral, uncommitted, agnostic light—that is, when we don't commit ourselves to it one way of the other in point of belief but take an endorsement-suspensive stand. The following would be an example:

If a woman is elected to the U.S. presidency in 2020, this will not alter the fact that most chief executives up to that time will have been male.

By contrast, a so-called *counterfactual* conditional has an antecedent that is belief contravening in contradicting accepted beliefs and will thereby be seen as false. For example:

If Paris were in Paraguay, then it would not be in Europe.

All answers to factual questions of the format:

What happens when . . . ?

are in a position to yield answers not only to questions of the factual type:

What would happen if . . . ?

but also questions of the counterfactual type:

What would have happened if . . . ?

In this light, the ordinary (factual) conditional

Since that ice cube did not melt, the temperature stayed below 32°F

and even the speculatively agnostic

If that ice cube did not melt (and no one is saying whether or not it did), then the temperature stayed below 32°F

contrast sharply with the counterfactual:

If that ice cube had not melted (which it did), then the temperature would have (or must have) stayed below 30°F.

All three of these conditionals pivot on the fact that ice melts at temperatures above 32°F. But they differ as regards the thesis "That ice cube did not melt," with the first taking it as true, the second agnostically viewing its truth status as unknown, and the third taking it to be false.

The pivotal feature of counterfactual conditionals is that they are belief contravening—that in the context of our prevailing beliefs, B, their antecedent (say, p) entails a contradiction:

$(p + B) \vdash$ (contradiction).

Counterfactual conditionals thus pivot on suppositions that are seen as false along the lines of "If Napoleon had stayed on Elba, the battle of Waterloo would never have been fought." Such counterfactuals *purport to elicit a consequence from an antecedent that is a belief-contradicting supposition, on evidence that it conflicts with the totality of what we take ourselves to know.*[2]

Supposition is, of course, a commonplace device that operates via such familiar locutions as "suppose," "assume," "what if," "let it be that," "consider the hypothesis that," and the like. A supposition is not an acknowledged fact, but a thesis that is accepted "provisionally" or laid down "for the time being"; it must be deemed false or at least uncertain to some extent—after all, if it were deemed true there would be nothing assumptive about it.[3] And it is the occurrence among the premises of an argument of such a suppositional hypothesis that renders the argument a "hypothetical."

In the final analysis, it is by no means necessary that with counterfactuals the antecedent's hypothesis represents an actual falsehood. What alone matters is that it clashes with what is believed to be true. A hypothesis is not rendered belief contradicting through failure to square with the actual facts, but simply and solely through a conflict with other propositions that are *accepted* as true in the circumstances at hand. In this regard the expression "counterfactual" carries misleading implications, although we shall continue to conform to the accustomed usage.

Counterfactual conditionals thus project contradiction-engendering antecedents into an environing manifold of accepted belief: they involve *conflicts* with our beliefs and not mere *supplementations* to them. The result is a self-contradictory mess. And the usual ground rules of conditionalizaion are cast to the winds on this basis. Instrumentalities need to be figured to bring order into chaos here.

This sort of thing is closely connected with our topic since conditionals effectively summarize the result of hypothetical inferences. A specifically "counterfactual conditional" is, in effect, nothing other than a conditional that claims a consequence for an antecedent that is a belief-contradicting hypothesis.

Counterfactuals have been a mainstay of speculative thinking for a long time. The Greek historian Herodotus (b. ca. 480 BCE) reasoned as follows:

If the Etesian winds were the cause of the Nile's annual flooding, then other rivers would be affected in the same way (which they are not).[4]

Hypothetical inferences of the sort at issue with counterfactual conditionals arise over a wide range of subject matter, which include:

1. Thought experimentation in tracing through the consequences of a disbelieved proposition (this occurs both in common life and in technical contexts).

2. Explanation of how things work in general—even under conditionals not actually realized.

3. The didactic use of hypotheses in learning situations.

4. Contingency planning (in everyday life).

5. Games and make-believe of all kinds.

6. *Reductio ad absurdum* reasoning and especially *per impossibile* proof (in logic and in mathematics).

Counterfactual inference accordingly plays a prominent part in both scientific and in common-life reasonings, and arises over an enormously wide spectrum of applications, ranging from the most serious to the most frivolous of contexts.

2 Types of Counterfactuals

Counterfactuals can be divided into the truthifying and the falsifying, which answer, respectively, to the format:

If the antecedent (which is deemed false) were true, then the consequent would obtain (as true)

and

If the antecedent (which is deemed true) were false, then the consequent would obtain (as true).

An example of the former (truthifying) type is:

If New York City were in France, then it would not be in North America.

And a counterfactual of the second (falsifying) type is:

If New York City were not in the United States, then Brooklyn would not be in the United States.

Counterfactual conditionals can also be classified in another way. A *strong* counterfactual is one where both antecedent and consequent are deemed to be unrealized:

If he had come into the room (which he didn't), then she would have left it (which she didn't).

By contrast, consider:

If he had come into the room (which he didn't), then she would have left it anyway (as she did).

This is a *weak* counterfactual whose consequent is seen as actually realized. (The semantic marker for such a counterfactual is "still" or "anyway.") The enthymematic basis for the two last-stated conditionals could be the belief that "She never remains in the same room with him." In conjunction with the antecedent this yields the consequent of the conditional as a deductive consequence.

Note too that both weak and strong counterfactuals can have modal consequents:

If he were a year older, then he could vote (which he can't). (Strong)

Even if there had been one lifeboat fewer, it would (still) have been possible to save all the passengers. (Weak)

It is important to note that even when their false antecedent is one and the same there can arise different counterfactuals that different counterfactuals answer different questions. Thus contrast:

If I had lots of money, I could [given what money can do] go to the racetrack and play the ponies

and

Even if I had lots of money, I would certainly not [given my frugal disposition] go to the racetrack and play the ponies.

So far there is no conflict here. These different counterfactuals answer different questions:

If you had lots of money, what *could* you do?

If you had lots of money, what *would* you do?

And here as elsewhere we must expect that when one asks different questions one will get different answers.

3 Belief-Contradicting Hypotheses and Burley's Principle

When we introduce an incompatible assumption into the manifold of our beliefs we must, of course, revise that family if consistency is to be maintained.

The reason for this situation lies in the logical principle of *the systemic integrity of fact*. For let it be that we accept p_1. Then let p_2 be some other claim that we flatly reject—one that is such that we accept $\sim p_2$. Now since we accept p_1, we will certainly also accept $p_1 \vee p_2$. But now consider the group of accepted theses: p_1, $p_1 \vee p_2$, $\sim p_2$. When we drop p_1 here and insert $\sim p_1$ in its place we obtain $\sim p_1$, $p_1 \vee p_2$, $\sim p_2$. And this group is still inconsistent. The long and short of it is that the facts purported by our beliefs are always united in a pervasively redundant structure. Every determinable fact is so severely hemmed in by others that even when we erase it, it can always be restored on the basis of what remains. Consider a graphic illustration of this circumstance. Suppose that x is located in a tic-tac-toe grid as follows:

Here we have the following facts:

(1) There is an x in the configuration.

(2) x is not in the first row.

(3) x is not in the third row.

(4) x is not in the second column.

(5) x is not in the third column.

(6) x is not on the diagonal.

(7) x is not at column-row position (3, 2).

Let it be that we erase one of the facts, say (5). Then, as we have already noted, the other contextual facts suffice to let us recover (5) straightaway by logical inference. But now suppose that we do not simply lose sight of (5) but actually *change* it, replacing it with something that entails the falsity of (5). Then, of course, we would also be obliged go on to deny either (6) or (7). The manifold of truth is *logically unified*: any change in one fact will always compel further changes in other facts.

As this example illustrates, and philosophers have long insisted, the fabric of fact is woven tight. G. W. Leibniz put the matter as follows:

There is nothing in the world of created beings whose perfected conception does not involve the conception of every other thing in the universe, since each thing influences all the others in such a way that, supposing that it were abandoned or changed, all things in the world would henceforward be different from the way they actually are. (To De Volder, 6 July 1701: in Leibniz 1879, p. 226; my translation)

The basic idea at issue here goes back (at least) to neo-Platonism.

The systemic density of facts means that they are so closely intermeshed with each other as to form a connected network. Any change anywhere has reverberations everywhere. In his influential *Treatise on Obligations*[5] the medieval scholastic philosopher Walter Burley (ca. 1275 to ca. 1345) laid down the rule—let us call it Burley's Principle—to the effect that: *When a false contingent proposition is posited, one can prove any false proposition that is compatible with it.* His reasoning was as follows. Let the facts be that:

p: You are not in Rome.

q: You are not a bishop.

And given *p* we will also have:

r: You are not in Rome or you are a bishop. (*p* or not-*q*)

All of these, so we suppose, are true. Let us now posit by way of a (false) supposition that:

Not-*p*: You are in Rome.

Obviously *p* must now be abandoned—"by hypothesis." But nevertheless from *r* and not-*p* we obtain:

Not-*q*: You are a bishop.

And in view of thesis *q*, this is, of course, false. We have thus obtained not-*q* by cogent inference from acknowledged truths—where *q* can be *any arbitrary true proposition.* And it is clear that this situation obtains in general. For let *p* and *q* be any two (arbitrary but nonequivalent) facts. Then all of the following facts will also of course obtain: $\sim(\sim p)$, $p \& q$, $p \vee q$, $p \vee (\sim q \vee r)$, $\sim p \vee q$, $\sim(\sim p \& q)$, and so on. Let us focus on just three of these available facts:

(1) *p*

(2) *q*

(3) $\sim(\sim p \,\& \,q)$ or equivalently, $p \lor \sim q$

Now let it be that we are going to suppose not-p. Then of course we must remove (1) from the list of accepted facts and substitute:

(1') $\sim p$

But there is now no stopping. For together with (3) this new item at once yields $\sim q$, contrary to (2). Thus that supposition of ours that runs contrary to accepted fact (viz., not-p) has the direct consequence that *any other arbitrary truth must also be abandoned.*

In making the counterfactual assumption of p where we in fact accept not-p, we suspend (or *bracket*, in Husserl's terminology) our actual belief to the contrary. We undertake a literal "suspension of disbelief." But the problem is how this belief-disruptive bracketing is to be reconciled with the rest of our environing beliefs. Burley's Principle ensures that this can never be a logically straightforward process. The systemic integrity of facts stands in the way. The threads of connection that pervade the fabric of truth resist and problematize the insertion of falsehoods.[6]

4 Validating Counterfactuals

The difficulty engendered by a belief-contravening hypothesis is inconsistency. And in view of this the only cost-free conditionals one can engender in the face of such a hypothesis are those that are informatively vacuous. To get beyond this, additional machinery must be introduced to bring order out of the chaos created by Burley's Principle.

The situation of counterfactual conditionals is unique. With factual conditionals (since-conditionals) and even with agnostically subjunctive conditionals ("If he were to come, then ...") we can simply adjoin the antecedent to what we know and proceed inferentially from there. But with counterfactual conditionals that antecedent is belief-contradicting: it stands in logical conflict with our stock of (putative) knowledge. Straightening out the mess that results must in the end call for novel and somewhat drastic measures by way of a revision of what we take to be true that involves an abandonment of part of it.

Nevertheless, which counterfactuals can be appropriately asserted will largely depend on what the real facts of the situation are. If I have five dollars in my pocket—but only then—will we have the counterfactual "If I had one more dollar in my pocket than I do, I would have (exactly) six dollars there." Only against a suitable background of accepted facts can counterfactual conditionals be sensibly maintained. Their semantical footing is not in alternative, inexistent worlds but in the real world—or what we accept as such.

It is helpful to adopt a bit of formalism. Let us employ the convention that a counterfactual conditional is to be represented as

$p\{B\} \mapsto q$,

which may be read as: "If p were true, which we take not to be so—not-p being a member of the set of our pertinent beliefs B (so that $\sim p \in B$)—then q would be true."

The guiding idea of the present approach to such statements is that a counterfactual conditional holds when the belief-contradicting supposition at issue in its antecedent will yield its consequent as a deductively valid conclusion when supplemented by *some suitable supposition-compatible beliefs* that are available within the overall manifold of our relevant beliefs B. The epistemic situation that prevails with counterfactuals accordingly has the following generic structure:

(disbelieved antecedent + certain accepted beliefs) ⊢ consequent

This approach represents a *derivability construal of counterfactuals*. The basic idea at issue here goes back to the ideas of F. P. Ramsey as summarized in R. M. Chisholm's statement that a counterfactual conditional "can be reformulated as a statement stating that the consequent is entailed by [i.e., logically derivable from] the antecedent in conjunction with a previous stock of knowledge."[7]

For the sake of an example, consider the conditional:

If John (who is in Detroit) were in Los Angeles, then he would be in California.

This thesis is projected in a context where we believe:

(1)　John is in Detroit.

(2)　John is not in Los Angeles.

(3)　Anyone who is in Los Angeles is in California.

Now the antecedent of the conditional instructs us is to suppose not-(2), namely that John is in Los Angeles. We must then, of course, abandon not only (2) but also (1). But we can unproblematically retain (3), which, in combination with our assumption of "John is in Los Angeles," yields the consequent of our counterfactual.

Again, suppose, for the sake of a somewhat more complex example, that we know the following (redundantly stated) facts:

(1)   Tom is among the present.

(2)   Bob is among the present.

(3)   Both Tom and Bob are among the present.

(4)   Tom and Bob apart, there are $n - 2$ people present.

And now let it be that we are to undertake the supposition:

Assume not-(3), namely that one of the pair Tom–Bob were not present.

To achieve consistency in the face of this assumption, we must abandon not only (3) but also either (1) or (2), the choice being indifferent. But (4) can be kept in place either way. So in point of acceptance/rejection the counterfactual assumption at issue confronts us with two alternatives:

(1), (4)/(2), (3)

(2), (4)/(1), (3)

Yet either way we would arrive at: "There are $n - 1$ people present." And on this basis we at once validate the counterfactual:

If either Tom or Bob were absent, then there would be $n - 1$ people present.

As such examples indicate, our approach to the issue of tenability is such that a counterfactual of the form

$$p \{B\} \rightarrow q$$

is acceptable when q is derivable from the combination of q with some appropriately contrived subselection of B-members. The problem is to determine which ones; for this doxastic, belief-derivability-geared approach to counterfactuals has its difficulties unless suitably elaborated and qualified. Consider, for example, the situation where we believe (among other things) the following three propositions:

(1) p

(2) $p \vee q$

(3) $p \vee \sim q$

And now suppose that we admit the assumption: not-(1), that is, $\sim p$. Together with (2) this yields q, and again with (3) it yields $\sim q$. Obviously, this will hardly do if a coherent account of counterfactuals is to be possible.

Recall the distinction between factual, agnostic, and counterfactual conditionals, as illustrated by:

Since he is coming, we will meet.

If he is/were coming, we would meet.

If he had come, we would have met.

In each case, the antecedent is in a different epistemic condition, respectively:

$p \in B$

$p \notin B$ and $\sim p \notin B$

$\sim p \in B$

In the first case we already have p in B, and in the second we can add p to B without creating any problems. No difficulty there. But in case three we have to readjust B to accommodate p. And as Burley's Principle indicates, this will always involve problems characteristic of the situation of counterfactuals.

5   The Problem of Restoring Consistency

On this basis Burley's Principle has far-reaching implications. Given the interconnectedness of facts, any and all fact-contradicting assumptions are pervasively destabilizing. As far as the logic of the situation is concerned, you cannot change anything in the domain of fact without endangering everything. Once you embark on a contrary-to-fact assumption, then as far as pure logic is concerned, all the usual bets are off. Changing one fact always requires changing others as well: the fabric of facts is an integral unit, a harmonious system where nothing can be altered without affecting something else.

Alterations in the manifold of facts lead us to embark on a process that potentially has no end. Thus suppose that we make only a very small alteration in the descriptive composition of the real, say, by adding one pebble to the river bank. But which pebble? Where are we to get it and what are we to put in its place? And where are we to put the air or the water that this new pebble displaces? And when we put that material in a new spot, just how are we to make room for it? And how are we to make room for the so-displaced material? Moreover, the region within six inches of the new pebble used to hold n pebbles. It now holds $n + 1$. Of which region are we to say that it holds $N - 1$? If it is that region yonder, then how did the pebble get here from there? By a miraculous instantaneous transport? Perhaps by a little boy picking it up and throwing it—but then which little boy? And how did he get there? And if he threw it, then what happened to the air that his throw displaced which would otherwise have gone undisturbed? Here problems arise without end. Every hypothetical change in the physical makeup of the real sets in motion a vast cascade of physical changes either in the physical constitution of the real or in the laws of nature at large. For what about the structure of the envisioning electromagnetic, thermal, and gravitational fields? Just how are these to be preserved as was given the removal and/or shift of the pebbles? How is matter to be readjusted to preserve consistency here? Or are we to do so by changing the fundamental laws of physics?

If overall coherence is to be achieved in the wake of a belief-contradicting supposition, this requires various deletions from the body of preexisting belief.[8] In the wider context of all the prevailing beliefs, counterfactual hypotheses are always paradoxical because *every belief-contradicting hypothesis poses conflicts with other available beliefs— conflicts that will require further adjudication.* The reality of it is that the logico-conceptual interlinkage of our beliefs is such that belief-contradictory suppositions always function within a wider setting of accepted beliefs $p_1, p_2, \ldots, p_n$ of such a sort that when one of them, for simplicity say p_1, must be abandoned owing to a hypothetical endorsement of its negation, nevertheless the resulting group $\sim p_1, p_2, \ldots, p_n$ remains collectively inconsistent.

The systemic integrity of facts indicates that we cannot make hypothetical modifications in the makeup of the real without thereby destabilizing everything and raising an unending series of questions. And not only do *redistributions* raise problems but even mere *erasures*, mere cancellations do so as well; reality being as is they require redistributions to follow in their wake. If by hypothesis we zap that book on the shelf out of existence, then what is it that supports the others? And at what stage of its production did it first disappear? And if it just vanished a moment ago, then what of the law of the conservation of matter? And whence the material that is now in that book-denuded space? Once more we embark upon a complex and baffling journey.

In order to avert such difficulties there must be some further guidance as to which antecedent belief should be retained and which abandoned. Accordingly, R. M. Chisholm added the proviso that only some *suitable* beliefs are to be retained. However, he never managed to provide any satisfactory account for implementing this condition of suitability.[9] And so, if a workable derivability theory of counterfactuals is to be obtained, then special measures will have to be derived to remedy this deficiency by specifying just which beliefs are to be available for the deductive work at issue. The difficulty is that when a belief-contradicting hypothesis is projected, at least some of the environing beliefs must always be abandoned in the wake of the hypothetical assumption. And this elimination can be accomplished in different ways.

Thus suppose that tigers were canines. Of course, this hypothesis arises in a context in which we, of course, know each of the following:

(1)   Tigers are felines.

(2)   Tigers are not canines.

(3)   No felines are canines.

The assumption at hand explicitly amounts to not-(2). But what of those other two theses? Are we to alter the boundaries of the classification "felines" (and so drop (1) as well), or keep these boundaries the same and so countenance tigers as somehow combination canines–felines (thus dropping (3))? Obviously we must, in the interests of mere self-consistency, adopt one or the other of these procedures if logical paradox is to be avoided. There are always conflicting alternatives in such cases, and the assumption itself gives us no directions for effecting the choice that arises here. This problematic situation portends problems for the analysis of counterfactuals. How can they be resolved?

The systemic harmony of facts is a circumstance whose significance for and impact on the counterfactualist is difficult to exaggerate. It means that any hypothesis deemed counterfactual is pervasively belief destabilizing in its introduction of contradictions. Thus if we assume, say, "Napoleon died in 1921," we have to choose between keeping his age (and thus having him born in 1869) and keeping his birthday (and thus having him live to age 152). And nothing about the counterfactual assumption itself provides any guidance here. The situation in this regard is totally ambiguous and indeterminate: We are simply left in the dark about implementing the counterfactual assumption at issue. Again, consider the assumption stipulated by "Suppose George Washington were living today." Here, too, perplexity is created by an ambiguity. Are we to exempt the Father of our Country from the normal processes of decay and dissolution and to suppose that the reports of his death were "greatly exaggerated," or are we to retain his prior demise and think of him as somehow reincarnated among our contemporaries? We can pinpoint this ambiguity by asking if the statement "George Washington did not die in 1799" is or is not to follow from the supposition.

Similarly, contrast the "natural" counterfactual:

If wood conducted electricity, then this stick would conduct electricity [since it is made of wood],

with the "unnatural" counterfactual:

If wood conducted electricity, then this stick would not be made of wood [since it does not conduct electricity].

Once we introduce that belief-contravening hypothesis, different possibilities for readjustment arise. And since *alternative* outcomes are always possible in such cases, a logical analysis of the situation will not of itself be sufficient to eliminate the basic indeterminacy inherent in counterfactual situations. A putative if-then relationship does not qualify as authentic if, when p is not self-contradictory, it permits claiming both $p \rightarrow q$ and $p \rightarrow r$ for logically incompatible q and r.

If counterfactual assumptions are to give rise to informative conditionals, such discrepancies must be eliminated. However, no purely formal or logical resources will resolve the issue of such discrepant counterfactuals; some logical-external, "material" or substantive mechanisms have to be introduced. Whenever an inconsistent group of theses confronts us—say, $p \vee r$, and $\sim p$ and $\sim r$—what logic can tell us is that at least one of them must be abandoned. But the question of which way to go is something that considerations of logic and abstract theory cannot provide. It emerges that counterfactual reasoning, like inductive reasoning, is ultimately a primarily functional project whose management requires pragmatic resources.[10] And at this point we need to turn from theoretical logic to applied epistemology.[11]

Yet is it not a defect of such a belief-consistency reductive approach to counterfactuals that purely logical (i.e., deductively logical) considerations alone cannot tell us what counterfactuals are acceptable in the absence of epistemological background information? Is this not a fatal flaw for the theory?

By no means! In factual and counterfactual matters alike the power of abstract reasoning alone is limited. Logic as such does not tell us what propositions are true but only what inferences are valid—and thereby what we have to accept as true *if* certain statements (the premises) are true. Analogously, counterfactuals will emerge as correct (i.e., tenable as truths), only *if* certain statements (the given enthymematic bases) are seen as salient truths. In either case, such truth-determination as there is is basis relative, conditionalized upon substantive decisions (to accept as true or to accept as salient, respectively) that lie beyond the power of logic alone. This circumstance is unavoidably inherent in the very nature of the altered situation that arises when we turn from factual to fact-controveningly suppositional matters.

8   Salience and Questioner's Prerogative

1 Salience and Formulation Sensitivity

A counterfactual conditional offers a response to a "what-if" question that contemplates a fact-contradicting circumstance. And in view of the inherent redundancy of fact-affirming propositions as per Burley's Principle, such a fact-abrogating supposition always requires additional disambiguation because of the systemic redundancy of truth. This means that belief-contradictory hypotheses—and hence counterfactual conditionals—will have to be handled with special care. For if manifold of the beliefs at stake were not radically restricted, the process of disambiguation could never be brought to an end, continuing *ad infinitum* because further issues could always be raised. Thus a workable approach to the consequences of a belief-contradicting hypothesis will have to focus not on the unmanageably wide range of *all* the abstractly available putative truths that are somehow relevant to the assumption at issue, but rather on the narrower, more restricted, and thus more manageable range of specifically *salient* beliefs that are deemed to be of special significance in the particular concrete setting of the given deliberations.

Consider the following illustrative example. We are given the number array:

1 2 2 1

And let us ask the counterfactual question: if that initial 1 were 2, what integer would then stand in second place?

We may suppose that the following propositions are salient in the problem context at hand:

(0) That four-integer group contains nothing other than 1s and 2s.

(1) The initial integer is a 1 and not a 2.

Table 8.1
Truth-Value Alternatives

Possible Array	Truth-Status of		
	(2)	(3)	(4)
2111		F	F
2112			
2121		F	
2122			F
2211		F	
2212			F
2221		F	F
2222	F		F

Note: All the entries are T (true) unless otherwise indicated.

(2)   That four-integer group is not uniform (i.e., is not all 1s and all 2s).

(3)   The same integer is at place 4 as at place 1.

(4)   That four-integer group contains the same number of 1s and 2s.

Let us now assume that not-(1), so that that initial integer is a 2 and not a 1. Since we suppose the problem situation to be such that (0) is fixity, there remain eight possibilities with respect to (2) through (4), as per table 8.1. As this shows, there is only one possibility—2, 1, 1, 2 to be specific—that leaves (2) through (4) intact. So relative to the indicated manifold of salient beliefs we arrive at

If that initial integer were a 2, then a 1 would occur at the second place.

To be sure, if different propositions were classed as salient—for example "There is a 2 in the second place"—then different results could follow.

Salience is a matter of belief evaluation, of informative normativity: The significance at issue with salience is a matter of comparative centrality: a belief is *salient* in the setting of a given problem-context when it is seen as being an item of knowledge that is not just relevant to but important for the range of questions and issues at hand. Salience is a matter not of substantiating the results of our inquiries but rather one of defining what the issues are that these inquiries address. Salience thus hinges on *questioner's prerogative*—it is a function of just exactly what issue is being put on the agenda. It is a matter of questions, not answers, and it pivots on achieving clarity about the issues on the agenda of deliberation. Certain facts about Napoleon will be salient in relation to his merits as a gen-

eral and other quite different facts about him will be salient in relation to
his merits as a husband. Accordingly, salience is context variable; the salient propositions serve to define the problem setting of an inquiry.

2 Salience Determination

The decisive role of salience in the management of belief-contradicting
suppositions comes into clear view when we confront the following tic-
tac-toe setup:

Given this diagrammatic representation there are various descriptively
different ways of articulating the beliefs that fix x's position here, specifically including the following three:

I	II	III
(1) At column/row position $(2, 1)$	(2) Not in rows 2 or 3 (3) Not in columns 1 or 3 (6) Not at column/row position $(2, 3)$	(4) Not on a diagonal (5) Not in row 2

All of these belief-sets convey the same informative result: all three fix
x's position at column/row position $(2, 1)$. All three structure the same
information differently. However, which of these belief groups is counted
as salient will make a significant difference for counterfactual analysis.
To see how this comes about, let us assume not-(1), and examine the consequences. Where would x be were it not at column/row position $(2, 1)$?

   With belief-set I we are left with nothing after making this assumption.
Not-(1) apart, there is nothing we can say here. And with belief-set II we
have to abandon one of theses (2) or (3) while nevertheless retaining the
other. We thus arrive at the result:

x is at one of: $(1, 1)$, $(2, 2)$, $(2, 3)$, $(3, 1)$.

On this basis $(1, 2)$, $(1, 3)$, $(3, 2)$, and $(3, 3)$ are out of the running in addition to $(2, 1)$. By contrast, with belief-set III we have to abandon one or
(4), (5), (6), and accordingly arrive at:

x is anyplace except $(2, 1)$ and $(2, 2)$.

Clearly these alternative results are different. And so in the context of counterfactual deliberations we have it that *saliency matters*. Even when "one and the same situation" is characterized in terms of different sets of salient beliefs we can arrive at different counterfactual conditionals. Differently organized (even though deductively equivalent) epistemic bases provide for different counterfactual conditionals.

In sum, the comparative significance and contextual centrality that we accord to our beliefs in view of the questions at issue will lead to different results in the wake of belief-contradictory suppositions. The questioner's prerogative prevails here—it is up to the questioner who puts a counterfactual issue on the agenda of consideration to specify what issues are on the agenda.

That counterfactuals are formulation sensitive with respect to salient beliefs has far-reaching consequences. For an important lesson comes to light, namely that the issue of which counterfactuals obtain as appropriate relative to a manifold of belief B is a function not only of the information contained in B (i.e., the propositions it entails), but also of the way in which this information is organized, in respect to issue relevancy as reflected by salience, so that: *counterfactual inference from a given body of information is formulation sensitive*. This complexity of having to heed the detailed way in which the relevant beliefs are highlighted in point of issue-pertinent salience is the price we pay for undertaking suppositions that fall afoul of our beliefs. The lesson is clear: To realize informative counterfactuals we must augment considerations of belief-acceptability as such by considerations of salience—of contextual significance.

One basic rule governing the concept of salience is:

If $p \vdash q$, and p is salient, then q cannot be salient.

This means we have to take a salient belief-set B as explicitly given: we cannot do "merely logical" supplementations to the set of salient beliefs —say, by adding $p \vee q$ when it already contains p. (In counterfactual contexts, belief-sets cannot be taken as belief *commitments* that are closed under deduction.)

In view of this rule, when p can be disaggregated into p_1 and p_2—so that $\vdash p \equiv (p_1 \& p_2)$—then by classing these two as salient we broaden the range of other propositions eligible for salience, since there will be various propositions that follow when p_1 and p_2 are conjoined into $p_1 \& p_2$ that do not follow from either p_1 or p_2 separately. Thus the more care we take to disaggregate a salient belief into greater detail the more elaborate our overall specifications of salience can become, and thereby our deliberations about counterfactual "what-if" issues can become more productive. For example, contrast the premise pair $\sim p$, $p \vee q$ with the pair $\sim p$, q. As far as deductive logic is concerned they combine to amount to the same thing, namely $\sim p \mathbin{\&} q$, and thus convey precisely the same information. But in the context of counterfactual reasoning they are very different. For suppose that we now project the assumption that p. In the case of the first pair, $\sim p$ must now be abandoned and $p \vee q$ becomes redundant in the wake of assuming p. Thus in point of post-assumption information, that first premise pair is simply annihilated. But in the second case, something substantial is still left in place, namely q. So here the conditional "Even if p, then (still) q" would be appropriate.

Salience, then, is a matter of configuring our overall beliefs to the situation of a particular problem-context, of effecting a concentration of focus within the overall manifold of our beliefs. And what is at work here is a matter of questioner's prerogative. It is the individual—whoever it may be—who puts the issue into consideration who has the privilege and bears the onus of specifying the question in such a way that questions of saliency are resolved. And there is inevitably something inquirer relative about this process of defining the question because the objective "facts of the matter" do not address issues of interest and concern: the facts of the situation do not themselves determine the focus of our questioning and the issues into which we can inquire.

A technical point must be noted. The question arises of whether the membership of the propositional set B of salient beliefs that are at issue with counterfactual conditionals needs to be explicitly spelled out as per $B = \{p_1, p_2, \ldots p_n\}$ or whether its membership can be descriptively indicates as per $B = \{\text{the set of beliefs I now hold}\}$. The answer here is that informatively tractable counterfactuals are predicated on an explicit access to the relevant beliefs. For, as will emerge below, the acceptability of such counterfactuals will hinge of the *comparative status* of those beliefs in point of informativeness, generality, lawfulness, and/or plausibility. And this is something one cannot assess without knowing specifically what those beliefs are.

But what if, at this point, we were to try to "objectify" matters by operating with the set of *all* truths (T) in place of the set of all issue-salient beliefs? This has two big problems: (1) There is no way to get there from here: we cannot come to grips with the totality of truths as such, having no way to get at the truth independently of our beliefs about it. (2) In view of Burley's Principle we have no practicable way to make the necessary consistency-restrictive readjustment once a falsehood is introduced into the comprehensive manifold of all truth.

Salience is even of utility in noncounterfactual contexts. I have a great many specific beliefs, say, $p_1, p_2, \ldots p_n$. But I am also inclined to the higher-level supposition—and perhaps even belief—that at least some one of those specific beliefs is incorrect. (Let this be s.) Clearly when s is compared to the totality of the p_i, an inconsistent set of beliefs results. But in any particular context of deliberation, some of those p_i may not be salient. Then, by seeing some of *these* as false, I save the consistency of the set. Moreover, in some—perhaps most—contexts, that higher-level supposition (s) may not be salient, so that I will have no difficulty affirming all of p_i, provided they are collectively consistent. Here too salience will depend on what the questions at issue are.

3 Changing the Range of Belief

The validation of counterfactuals is an information-sensitive issue. Increasing the information available in our beliefs might afford us offer ampler opportunities for validating further counterfactuals or, alternatively, it could destabilize existing patterns to an extent that unravels previously available ones. Let us consider some examples of this situation, beginning with an instance of gained counterfactuals with enlarged B. Suppose a situation in which we know all of the following:

Tom is 30 years old.

Tim is 31 years old.

Tod is 31 years old.

Tom, Bob, and Tod are all members of the club.

We now cannot validate:

If Tom were 31 years old, then all members of the club would be so.

But if we added the further datum

Tom, Tim, and Tod are the *only* members of the club,

then we could indeed validate this counterfactual. Augmenting our body of B-available information here provides for further counterfactuals.

On the other hand, the failure of monotonicity in relation to counterfactual conditionals means that it is also possible that in amplifying B we might disestablish certain counterfactuals. The following situation illustrates this prospect of lost counterfactuals with enlarged B. Let it be that:

Tom, Tim, and Tod were the only members of the club.

Ted was not a member.

We now have:

If Ted had joined the club, then all members would (still) have names beginning with T.

But suppose we *also* know, additionally, that as a matter of fixed personal policy Ted never joins anything without his friend Bob. Then our counterfactual would be disestablished. Thus enlarging the manifold of our beliefs with additional information can not only expand the range of acceptable counterfactuals but can contract it as well, since preexisting patterns can then become unraveled.

Moreover, there is also the prospect of gained counterfactuals with diminished B. Consider the set-up:

x	

Here we have the following facts.

(0)   There is exactly one x.

(1) x is in column/row position $(1, 1)$.

(2) x is not in row 2.

(3) x is not in column 2.

And now assume that not-(1), by adopting the following supposition:

(S)   Not-(1), that is, x is not in column/row position $(1, 1)$.

Clearly, if we retain (0), either (2) or (3) must now be sacrificed. But seeing that we have no guidance whatever as to which, and all we can now maintain is the disjunctively indefinite conditional:

If x were not at $(1, 1)$, it would be at $(1, 2)$ or $(2, 1)$ or $(2, 2)$.

Thus what we have is the triviality that if x were not at $(1, 1)$ it would be elsewhere. Observe, however, that if we *impoverished* our stock of salient beliefs and lost sight of (3) altogether, and only had the rest at our disposal, then the loss of (1) would still leave (2) in place, so that our supposition (S) would, given (2), lead to the more definite result:

If x were not at $(1, 1)$, it would be at $(2, 1)$.

Here again, differences in saliency in the constitution of background knowledge prove to be crucial for the range of counterfactuals that can be validated.

Finally, the prospect of lost counterfactuals with diminished B is too obvious to need further substantiation. And so the overall lesson is clear: the exact range of the issue-salient beliefs—and in particular those deemed more fundamental and lawlike—is something that will be pivotal for the determination of circumstantially cogent counterfactuals.

4 Questioner's Prerogative

We have arrived at a point where the pivotal principle of the questioner's prerogative and the constitution of the agenda need to be clarified.

Consider the following example, due to W. V. O. Quine.[1] There are two possibilities:

If Bizet and Verdi had been fellow countrymen, then Bizet would have been Italian

or

If Bizet and Verdi had been fellow countrymen, then Verdi would have been French.

Given that France and Italy are different countries with different nationals, two further salient facts lie before us:

(1) Bizet was a Frenchman.

(2) Verdi was an Italian.

These two contentions are altogether on a par with one another so that the context-relevant considerations of priority are of no avail here: from the angle of informativeness, the available alternatives are symmetric and thereby indifferent. Here we can go no further by way of conclusion than to arrive at the inconclusive disjunction:

If Bizet and Verdi had been fellow countrymen, then either Bizet would have been Italian or Verdi would have been French.

To get beyond this, we would have to be more specific about the question at hand. There are several possibilities here:

If Bizet had been a countryman of Verdi, what nationality would they have been?

If Verdi had been a countryman of Bizet, what nationality would they have been?

If Bizet and Verdi were fellow countrymen, what nationality would they be?

With the first wording of the question we are, in effect, instructed to prioritize (2) over (1), while with the second wording the reverse is the case and (1) gains the priority. But in our original question the wording carefully avoided any prioritization, and for that very reason it permits no definite conclusion. In this case all that we can justifiably claim is the disjunctive conditional stated above.

The issue of questioner's prerogative is pivotal here. For the tenability of counterfactuals is not just a matter of available *information* but also one of different *questions* being addressed with different answers accordingly called for. The matter is one not of different contexts of belief but of different contexts of inquiry.[2]

Thus consider yet another example due to W. V. O. Quine:[3]

What if Julius Caesar had been the Allied commander in Korea?

To come to terms with this question consider some saliently relevant things we know about Julius Caesar and about the Korean conflict:

(1)   Julius Caesar died in 44 BCE.

(2)   The Korean War was fought in 1948–1953.

(3)   The most powerful weapons in the first century BCE were catapults.

(4)   The most powerful weapons in the 1950–1960 era were atomic bombs.

(5)   In warfare, Caesar would use the most powerful weapons available to him.

Now, in connecting Caesar and Korea, we have to make a basic choice. Either we carry Julius Caesar forward to the time of the Korean War (and abandon (1)) or we take the Korean War back to Julius Caesar's day (and abandon (2)).

In taking the former course, we would validate

If Julius Caesar had been the Allied commander in the Korean War in the twentieth century, then he would have used the atom bomb.

In the latter event we would validate

If Julius Caesar had been the Allied commander in a Korea-like war in the first century BCE, then he would have used catapults.

But of course the original question is wholly undecided between the two alternatives. And so the issue here is not "Which counterfactual answers our question appropriately?" but rather "Just what is it that this question is asking?" The tenability of the counterfactual at issue depends once again on just exactly what the question is to be answered by it.

Consider another example. Let it be that we accept both of the following contentions:

(1)   John F. Kennedy was assassinated.

(2)   Lee Harvey Oswald (and no one else) assassinated John F. Kennedy.

We can now project the following two counterfactuals:

If Oswald did not assassinate Kennedy, then [given that Kennedy was assassinated as per (1)] someone else did.

If Oswald had not assassinated Kennedy, then [given that he (and no one else) assassinated Kennedy as per (2)] Kennedy would not have been assassinated.

The first of these is basically an indicative conditional predicated on retaining (1) and subordinating (2) to this determination in the face of the Oswald-innocence hypothesis. The second is a more complex counterfactual conditional predicated on breaking (2) up into (2.1) Oswald assassinated Kennedy, and (2.2) No one other than Oswald assassinated Kennedy. And now, given that our hypothesis demands rejection of (2.1), we note that not-(2.1) plus the still-retainable (2.2) yields not-(1).

Both of these counterfactuals are plausible as they stand—but only with different contextual beliefs accorded priority. For they respond to distinctly different questions, namely:

If Oswald did not assassinate Kennedy, then who did?

Had Oswald not assassinated Kennedy, how would Kennedy have fared?

As this example indicates, one seemingly selfsame counterfactual antecedent poses different issues in different contexts. And, of course, we must be prepared for the upshot that different questions have different answers.[4]

5   Additional Problems

Counterfactual analysis turns on consistency restoration through belief reduction, by ruling some otherwise acceptable beliefs out of the range of consideration. Considerations of salience certainly help to accomplish this end. But unfortunately we must now acknowledge the somewhat dismaying fact that saliency considerations will often not of themselves suffice validate informative counterfactuals. Still stronger measures for reducing our belief commitments may yet be required—further eliminative measures that make some salient beliefs give way to others for the sake of restoring consistency.

Consider the counterfactual question:

What would have happened if he had driven faster?

We may suppose that the background of salient belief is as follows here:

(1)   He averaged 80 mph for the trip.

(2)   In going faster he would cover the same distance in less time.

(3)   He traveled on busy roads.

(4)   Driving faster than 80 mph on busy roads he would have had an accident that precludes his reaching his destination.

Let us now adopt the hypothesis

(0)   He drove faster than an average of 80 mph.

This, of course, requires abandoning (1) for the sake of consistency. But the following two further options shall remain as regards retention/rejection:

(0), (2), (3)/(4)

(0), (3), (4)/(2)

We are thus presented with a choice between (2) and (4)—that is, between the counterfactuals:

If he had driven faster, he would have arrived in less time

and

If he had driven faster, he would have had an accident and so not arrived at all.

Supposing (as we may) that there is nothing further at hand to guide a choice between (2) and (4), the best we can do here is to arrive at the *disjunctive* conditional:

If he had driven faster he would either have reached his destination earlier or not have reached it at all because of an accident.

When counterfactual assumptions confront us with an indifferent choice between distinct alternatives, the best we can do is take refuge in the security of disjunction.[5] Indecision and indefiniteness here exact the price of vacuity.

And so, while saliency is critically important for hypothetical reasoning, it is still not the end of the matter. Even after the totality of relevant belief commitments is reduced to the diminished subset of issue-salient beliefs, the result will, in general, still be incompatible with our counterfactual hypothesis in a way that forces us into disjunctive vacuity. To realize an informatively more definite resolution in such cases we must reduce the set B of issue-salient beliefs yet further to obtain a supposition-compatible basis for reasoning. This issue will be addressed in the next chapter.

6 Methodological Reflections

At this point the overall structure of the present approach to counterfactual validation should be clear.

The starting point of the analysis lies in acknowledging the impetus of Burley's Principle that truth is a tightly woven fabric and that a hypothetical change at any one place has ramifications throughout. There is no way to integrate a falsehood into the manifold of truth—or even of our commitments in point of putative truth—without a pervasive revision that reaches into virtually every corner.

To introduce a suppositional falsehood into the manifold of accepted truth leads to destructive ramifications that resist confinement to convenient limits. When p is the case, all that abstract logic enables us to do is to affirm such trivialities as:

If not-p were the case, then all the things that follow from not-p would be.

If not-p were the case, then many facts would become unraveled.

If not-p were the case, then matters would be otherwise in point of p and some of its implications.

If not-p were the case, then many facts would stand differently.

And so on. Such limitations are all that mere logic enables us to resume from the wreckage of inconsistency induced by counterfactual assumption. A satisfactory theory must arrive at *informative* counterfactuals that go beyond the effectively vacuous confines of such trivializations. And to secure informatively substantial counterfactuals we must go well beyond the resource of mere logic.

When a what-if question of the sort that gives rise to counterfactuals is posed, the first step to be made is one that takes us from our mere beliefs about the truth to the greatly diminished submanifold of issue-salient belief. Our guiding principle here is the issue of constituting the question's agenda—of delineating the issues on which the inquiry is to focus. And at this point the issue of questioner's prerogative comes to the foreground. Through effecting a substantial reduction in our now overextended belief commitment, focusing on salient issues is a major step toward reducing counterfactuality-engendered chaos to meaningful proportions.

But it is not yet enough. Even with merely salient beliefs at hand inconsistencies among our acceptance-commitments may well yet remain. It is still all too readily possible that the rendered beliefs are too extensive by way of yet outrunning the limits of consistency.

When this happens our belief commitments need to be reduced further if we are to break out the cycle of inconsistency. Additional retrenchments will yet be needed with some of these salient beliefs giving way to others. How this is to be accomplished is the crucial issue to which we now turn.

First, however, one important lesson deserves stress. While the more familiar conditionals of factual purport require only a consideration of truth relationships, the step into the realm of supposition and counterfact calls for the additional machinery of saliency and retention prioritization. Merely semantic (i.e., truth-related) considerations suffice for the analysis of factual conditionals. But not for counterfactuals.

It thus deserves to be emphasized that this account of counterfactuals is *doxastic* (i.e., belief oriented) in its bearing rather than *semantic*: it is cast in terms of acceptability or assertibility conditions rather than truth conditions.[6] Thus it would, strictly speaking, be proper to speak of counterfactuals as being appropriate or correct rather than true. But the difference is somewhat academic, seeing that the acceptability at issue is in effect acceptability as true.

This feature of the centrality of questioner's prerogative—of specifying the content of the question being addressed—makes the present account of counterfactuals one that is geared to erotetic hermeneutics rather than one grounded in semantics alone.

9   On Validating Counterfactuals

1   Validating Counterfactuals

The definitive feature of counterfactual conditionals is that they are belief-contravening—that in the context of the overall manifold of our prevailing beliefs, **B**, their antecedent (say, p) entails a contradiction:

$(p + \textbf{B}) \vdash$ (contradiction).

With factual conditionals (since-conditionals), and even with agnostically subjunctive conditionals ("If he were to come, then ..."), we can simply adjoin the antecedent to what we know and proceed inferentially from there. But with counterfactual conditionals that antecedent is belief-contradicting: it stands in logical conflict with our stock of (putative) knowledge. Straightening out the mess that results must in the end call for novel and somewhat drastic measures by way of a revision of what we take to be true that involves an abandonment of part of it.

But nevertheless, which counterfactuals can appropriately be asserted will largely depend on what the real facts of the situation are. If I have five dollars in my pocket—but only then—will we have the counterfactual "If I had one more dollar in my pocket than I do, I would have (exactly) six dollars there." Only against a suitable background of accepted fact can counterfactual conditionals be sensibly maintained. Their semantical footing is not in alternative, inexistent worlds but in the real world—or what we accept as such.

Suppose, for the sake of a somewhat more complex example, that we know the following (redundantly stated) facts:

(1)   Tom is among the present.

(2)   Bob is among the present.

(3) Both Tom and Bob are among the present.

(4) Tom and Bob apart, there are $n - 2$ people present.

And now let it be that we are to assume not-(3), namely:

One of the pair Tom-Bob was not present.

To achieve consistency in the face of this assumption, we must abandon not only (3) but also either (1) or (2), the choice being indifferent. But (4) can be kept in place either way. So in point of acceptance/rejection the counterfactual assumption at issue confronts us with two alternatives:

(1), (4)/(2), (3)

(2), (4)/(1), (3)

Yet either way we would arrive at: "There are $n - 1$ people present." And on this basis we at once validate the counterfactual:

If either Tom or Bob were absent, then there would be $n - 1$ people present.

But how—if at all—would it be possible to get beyond this?

As already indicated, approach to the issue of tenability is such that a counterfactual of the form

$$p \{\mathbf{B}\} \!\!\rightarrow q$$

is acceptable when q is derivable from the combination of q with some appropriately contrived subselection of the set $\mathbf{B}$ of background beliefs. And the problem is to determine which ones. For we have to readjust B to accommodate p. And just this will inevitably involve problems of acceptability that are unique to the situation of counterfactuals.

Whenever a *belief-compatible* assumption is made, one can simply add this to our stock of beliefs and proceed to draw deductive inferences in the usual manner—and it is exactly in this way that substantive conditionals are validated. But with *belief-incompatible* suppositions, the situation is drastically transformed. For here—as we have seen—Burley's Principle means that one always faces a plurality of incompatible results according as that ("counterfactual") hypothesis is supplemented with different preexisting beliefs. For when an assumption p is belief contradictory relative to a set $\mathbf{B}$ of issue-salient beliefs, there will always be (at least) two different belief-subsets of $\mathbf{B}$, namely B_1 and B_2, such that

$(p + B_1) \vdash q_1$

$(p + B_2) \vdash q_2$

where q_1 and q_2 are incompatible. The various different ways of reconciling an incompatible p with **B** lead to logically incompatible results. The logical coherence of consequencehood is lost. How is it to be restored?

2   Systemic Fundamentality and the Tenability of Counterfactuals

Reconstituting the manifold of salient belief in the wake of a belief-contravening supposition is an exercise in damage control through cognitive economy. The items of putative knowledge that constitute that belief manifold have not come free of charge—they are hard-won fruits of inquiry through which we endeavor to answer our questions about the nature of things. Faced with a conflicting assumption we naturally want to salvage as much information as we can—to maximize the epistemic value of what we are able to preserve. And what comes to our aid here is the impetus to systematicity and generality that in inherent in the quest for informative utility.

The crux of this particular sector of cognitive economy lies in a principle of the conservation of information, which calls for prioritizing our beliefs in point of generality of scope and fundamentality of bearing. To see how this comes into operation consider the counterfactual:

If he had been born in 1999, then Julius Caesar would not have died in 44 BCE but would be a mere infant in 2001.

This arises in the context of the following issue-salient beliefs:

(1)   Julius Caesar was born in 100 BCE.

(2)   Julius Caesar is long dead, having died at the age of 56 in 44 BCE.

(3)   Julius Caesar was not born in 1999 CE.

(4)   Anyone born in 1999 CE will only be an infant by 2001.

(5)   People cannot die before they are born.

And let us now introduce the supposition of not-(3) via the following assumption:

Julius Caesar was born in 1999 CE.

In the face of this assumption we must, of course, follow its explicit instruction to dismiss (1) and (3). But (4) is safe, inherent in the very definition of infancy. Yet even then the residual inconsistency will still confront us with two distinct acceptance/rejection alternatives:

(2), (4)/(1), (3), (5)
(4), (5)/(1), (2), (3)

In effect we are constrained to a choice between (2) on the one hand and (5) on the other. At this point, however, the "natural" resolution afforded by impetus to informative systematicity has us prioritize the more general (5) over the case-specific (2), effectively eliminating the first alternative. The conclusion of the initial counterfactual then follows at once.

The process at issue here can also be illustrated vividly by another example from the literature, namely:

If this rubber band were made of copper, it would conduct electricity.

The counterfactual question that confronts us here is:

If this rubber band were made of copper, what then?

And this question arises in an epistemic context where the following beliefs are salient:

(1)   This band is made of rubber.

(2)   This band is not made of copper.

(3)   This band does not conduct electricity.

(4)   Things made of rubber do not conduct electricity.

(5)   Things made of copper do conduct electricity.

Against this background we are instructed to accept the hypothesis

Not-(2): This band is made of copper.

The following two propositional sets are the hypothesis-compatible maximal consistent subsets of our specified belief-set B:

{(3), (4)}   corresponding to the acceptance/rejection alternative (3), (4)/
             (1), (2), (5)

{(4), (5)}   corresponding to the acceptance/rejection alternative (4), (5)/
             (1), (2), (3)

The first alternative corresponds to the counterfactual

If this band were made of copper, then copper would not conduct electricity [since this band does not conduct electricity].

And the second alternative corresponds to the counterfactual

If this band were made of copper, then it would conduct electricity [since copper conducts electricity].

There is clearly a problem of cotenability here: to all appearances these conditionals lead us from one selfsame hypothesis to logically incompatible conclusions. We cannot have it both ways. So, in effect, we are driven to a choice between (3) and (5), that is, between a particular feature of this band and a general fact about copper things.

And here considerations of systemic informativeness come to the rescue. For its greater generality qualifies (5) as systemically more informative, and it is just this that renders its prioritization appropriate in the present circumstances. Accordingly, we will retain (4) and (5) along with not-(2), and therefore accept that second counterfactual as correct and abandon the first. In the wake of counterfactual suppositions we want to use damage control and save as much as we can; systemic informativeness functions as the pivotal factor in this regard.

3   A Derivability Theory of Counterfactuals

Counterfactual conditionals thus pivot on suppositions that are seen as false along the lines of "If Napoleon had stayed on Elba, the battle of Waterloo would never have been fought." Such counterfactuals *purport to elicit a consequence from an antecedent that is a belief-contradicting supposition, one that conflicts with the totality of what we take ourselves to know*.[1] And the guiding idea of the present approach to such statements is that a counterfactual conditional holds when the belief-contradicting supposition at issue in its antecedent will yield its consequent as a deductively valid conclusion when supplemented by *some suitable supposition-compatible beliefs* that are available within the overall manifold of our relevant beliefs **B**. The epistemic situation that prevails with counterfactuals accordingly has the following generic structure:

(disbelieved antecedent + certain accepted beliefs) ⊦ consequent

This general line of approach represents a *derivability construal of counterfactuals*. The basic idea at issue here goes back to the ideas of F. P.

Ramsey as summarized in R. M. Chisholm's statement that a counterfactual conditional "can be reformulated as a statement stating that the consequent is entailed by [i.e., logically derivable from] the antecedent in conjunction with a previous stock of knowledge."[2]

A counterfactual of the form "If p obtained (which it does not), then q would be the case" will here be symbolized as:

$$p \{\mathbf{B}\} \mapsto q$$

where $\mathbf{B}$ is the wider context of relevant beliefs. However, for matters to work out, we have to curtail $\mathbf{B}$ so as to render its reduced version appropriately compatible with p. Specifically, with a suitably p-adjusted streamlining of $\mathbf{B}$ in hand we can indeed validate a conclusion as per

$$(p + \mathbf{B}/p) \vdash C \text{ or (abbreviatedly) } p\{\mathbf{B}\} \mapsto C,$$

where $\sim p \in \mathbf{B}/p$.

For the sake of an example, consider the conditional:

If John (who is in Detroit) were in Los Angeles, then he would be in California.

This thesis is projected in a context where we believe:

(1)   John is in Detroit.

(2)   John is not in Los Angeles.

(3)   Anyone who is in Los Angeles is in California.

Now the antecedent of the conditional instructs us to suppose not-(2), namely that John is in Los Angeles. We must then, of course, abandon not only (2) but also (1). But we can unproblematically retain (3), which, in combination with our assumption of "John is in Los Angeles" yields the consequent of our counterfactual.

The pivotal idea here has it that counterfactual entailment is encapsulated in the principle that $p \{\mathbf{B}\} \mapsto q$ is to be validated as obtaining on the following basis:

1. Form the set $\mathbf{B}$ consisting of all p-relevantly salient beliefs.

2. Substitute $\sim p$ for p in this set so as to obtain an overall inconsistent set.

3. In this revised set, identify that consistent subset $\mathbf{B}/p$ (or, if several, then all those consistent subsets) which breaks the chain of inconsistency

at the weakest link in point of counterfactual plausibility (i.e. systemic informativeness).

4. Verify that q is a consequence of $\mathbf{B}/p$ (in every case).

In sum, it can be said that $p \{\mathbf{B}\} \mapsto q$ obtains iff

$(p + \mathbf{B}/p) \vdash q$ for all p-suitable subsets $\mathbf{B}/p$ of $\mathbf{B}$,

where p-suitability is a long story, shortly to be told.

4 Fundamentality Prioritization and Retention Precedence

When we deal with an implication $p \Rightarrow q$ in ordinary factual situations we need never look beyond p and q themselves to determine the tenability of this relationship. But in counterfactual situations, something else becomes critical, something that "does not meet the eye," namely the belief context $\mathbf{B}$. Counterfactuals become a matter of restoring consistency in the optimal way in point of retention prioritization. And this optimization is a matter of eliminating the weakest links in our chain of commitment.

But just how is that systemically "weakest link" among the propositions at issue is to be determined? The answer here lies in the consideration that strength in this context is a matter of systemic informativeness—of generality, lawfulness, and range of application. On this basis, we have a "fundamentality preference" through which generalities gain precedence over like particularities and existence precedes essence. Systemic fundamentality is the key, so that in general those statements whose abandonment creates less deformation in an overall view of the facts gain precedence over those whose abandonment would create more.[3] Thus conceptual relations (definitions, for example) take priority over factual generalizations. But these in turn—be they laws or normalities—have priority over mere facts. And even these are not created equal, since norms of practice (for example) will advantage some facts over mere matters of brute contingency.

Display 1 specifies the priority order regarding the fundamentality of our beliefs in matters of counterfactual reasoning. These levels of fundamentality represent and prioritize four key categories for hypothetical thought: *Meaning, Existence, Law, and Fact* (MELF). This ordering is crucial for determining precedence in conflict-resolution in the wake of belief-contravening assumptions. It establishes an order of retention-eligibility that reflects fundamentality in the systemization of our knowl-

Display 1
The MELF Levels of Systemic Fundamentality (Retention Priorities)

(M)	Definitional/Taxonomic Beliefs ("Elephants are vertebrates")
(E)	General Ontic Beliefs ("Elephants exist")
(L)	Lawful/Nomic Beliefs ("Elephants are gray")
(F)	Item-specific Beliefs ("That creature is an elephant [or: is gray]")

edge regarding X's views prioritization in the following light: the terms of reference at issue, the X's and their taxonomy, how these items function at the level of generality, and how matters stand differently with respect to particular individuals.

This MELF ordering of *meaning, existence, lawfulness,* and *fact* is a matter of specifying first the vocabulary and basic terms of reference, then the objects of discussion, then the basic principles and rules of the domain, and then the particular facts that feature in the discussion. The priority relationships shown in display 1 thus reflect what are the *standing presumptions* built into the operative groundrules of counterfactual discourse. They indicate how precedence and prioritization work in the absence of case specific specifications to the contrary. What we have here, then, is a general principle of fundamentality prioritization—one whose key consequences include a more specific principle of generality preference of laws over particular descriptive claims. Considerations of nomic fundamentality are paramount here: lawfulness, commonality, normality and their congeners.[4] It is the pivotal role of systemic fundamentality in counterfactual reasoning that marks this as a matter less of abstract rationality and logic than of common sense.

By way of illustration, contrast the counterfactuals

If he had thrown the switch, the light would have gone on (because that's how the system is designed to work).

If he had thrown the switch nothing would have happened (because the connection is broken).

Of course if we *know* the connection is broken, the second counterfactual wins out. But absent such knowledge, the first prevails. And for good reason. For the issue is one of normalcy versus malfunction. Our beliefs are as follows:

(1)   He did not throw the switch.

(2)   The light did not go on.

(3)   Whenever the switch is thrown, the light goes on.

Assumption: not-(1): He threw the switch. To resolve consistency in the face of the assumption, either (2) or (3) must be jettisoned. And in prioritizing the lawful (3) over the merely factual (2), we best maintain the lawfully general, "normal" order of things.

Some theorists regard the issue of the maintenance of lawful order as a matter of "expectation" and regard the corresponding prioritization of conditionals that shift the course of events further from the reign of the expected as a matter of the comparative "mutability" of the corresponding beliefs. This approach to prioritization seems perfectly consonant with a theoretical analysis that sees the issue as one of epistemic fundamentality. For if people's de facto handling of counterfactuals is by and large in its conformity to the theoretically mandated properties of the communicative situation, then such comportment is only to be expected.

It should be stressed that the prioritization of lawfulness over mere fact in counterfactual contexts also serves to make normalcy into a key aspect of fundamentality. Thus consider the counterfactual

If I had not thrown the switch the lights would not have gone on.

The relevant beliefs are:

(1)   Unless there is a fluke (such as a glitch in the wiring) the lights go on when and only when the switch is thrown.

(2)   I threw the switch.

(3)   The lights went on.

(4)   The situation was normal—no glitch, no fluke.

In the wake of the supposition of not-(2)—that is, that I had not thrown the switch—retention of the lawful thesis is at issue. Namely, (1) forces a choice between

(I)   (4) dominates (3), and we have ". . . then the lights would not have gone on (because there was no glitch)," and

(II)   (3) dominates (4), and we have ". . . then there would have been a glitch and the lights would (still) have gone on."

As long as a principle of normality is in operation so that we prioritize the normal and dismiss glitches as remote abnormalities, alternative (I) will win out, notwithstanding (II)'s vote in favor of the status quo.

Or again consider the following situation, in which our beliefs are:

(1)   Our Z engine did not fail.

(2)   Our Z engine did not have a defective grommet.

(3)   90 percent of Z engine failures are due to defective grommets.

(4)   We have no grounds for seeing our engine as other than typical.

And we are to assume not-(1), that is: Suppose our engine had failed.
    Given this supposition, we have to choose between two alternatives:

(I)   retaining (3) and (4) alongside of not-(1), we conclude: "Our engine (most probably) did have a defective grommet"—and accordingly jettison (2); and

(II)   retaining (2) and (3) alongside of not-(1), we conclude: "Given that our engine failed despite a totally healthy grommet, we have to regard it as a nonstandard exemplar whose of its type and accordingly jettison (4)."

With (I) we effectively subordinate (2) to (4), and with (II) we effectively subordinate (4) to (2). The question becomes one of choice between a generalized normality presumption and a mere claim of specific fact. And here the prioritization of the MELF ordering in counterfactual reasoning rules in favor of the first alternative. Such examples once again serve to indicate that problematic counterfactuals can be accommodated within the framework of the present approach.

    The examples just considered illustrate a general state of affairs. As we saw above, in *factual* situations with candidates of the form $p\,[B]\!\!\mapsto q$ it is the *plausibility* of a belief that determines its comparative strength or weakness. However, in *counterfactual* situations, with conditionals of the form $p\,\{B\}\!\!\mapsto q$, it is not plausibility but *systemic fundamentality* that provides the pivot. That is to say, the more deep-rooted and fundamental a thesis is in its breadth and scope, the greater its strength in the relevant sense of the term. However, this loose idea needs to be pinned down in greater detail.

    What is thus crucial with counterfactuals is the determination of precedence and priority for right-of-way determination in acceptance/rejection for consistency-restoration in cases of conflict. And here we proceed on the basis of the rule that:

In counterfactual reasoning, the priority among the issue salient beliefs is determined in terms of their standing in the systemic context at hand.

The situation can be summarized in the unifying slogan that with counterfactuals we prioritize beliefs on the basis of *systematicity preference*. But this matter of strength/weakness is now determined with reference to informativeness within the wider context of our knowledge. When we play fast and loose with the world's facts we need the security of keeping its fundamentals in place. In particular, it is standard policy that *in counterfactual contexts, propositions viewed as comparatively more informative in the systemic context at hand will take priority over those which are less so.* While revisions by way of curtailment and abandonment in our family of relevant belief are unavoidable and inevitable in the face of belief-countervailing hypotheses, we want to give up as little as possible. And here the ruling principle is: "Break the chain of inconsistency at its weakest link in point of systemic informativeness."

Thus in dealing with belief-contravening suppositions the operative rule is to prefer those consistency-restoring alternative(s) that minimize the maximal plausibility/priority of the rejected propositions. This *minimax selection principle* represents the policy of "breaking the chain of inconsistency at its weakest link."

Some further illustrations are in order. Suppose we have the beliefs:

(1)   Elephants exist [level 2 ontic belief].

(2)   Elephants are not three-headed [level 3 lawful belief].

(3)   No species of animal is three-headed [level 3 lawful belief].

And now let it be that we suppose not-(2): Elephants are three-headed creatures. We would then have to choose between (1) and (3). And here the Principle of Fundamentality has us arrive at the counterfactual: "If elephants were three-headed creatures, then there would actually be three-headed creatures (since elephants exist)."

Again, consider the following situation, in which we have the following salient beliefs:

(1)   Elephants exist [level 2 general belief].

(2)   All elephants are gray [level 3 lawful belief].

And now let it be that we project the following supposition:

Suppose there were no gray creatures.

We then face a choice between abandoning (1) or (2). Two alternatives accordingly confront us:

(I)   To have (1) yield way to (2) and thus effectively endorse the counterfactual:
If there were no gray creatures, then elephants would not exist (because elephants are gray).

(II)   To have (2) yield way to (1) and thus effectively endorse the counterfactual:
If there existed no gray creatures, then elephants would not be gray (because elephants do exist).

In effect we are led to a choice between (1) and (2). And here the precedence order given in display 1 indicates that (1) predominates so that alternative (B) in fact prevails.

The question "Where is the weakest link in a chain of interrelated beliefs?"—"Which is the weakest thread in a web of belief?"—is ambiguous and can be construed in two significantly different senses, namely the weakest in point of *evidentiation* on the one hand, and on the other the weakest in point of what might be called *systemic centrality*.

Now as regards evidentiation, every belief stands on its own—the question is simply: how well does the available information substantiate it in any specific case? But as regards systemic centrality, the question is one of the *types* of contention at issue and relates not to the specific information that it conveys but to the generic mode and manner in which it conveys information. It is a matter not of the specific and particular content of a belief but of the generic informativeness of the sort of contention at issue. Here we proceed not inductively but categorically.

The difference that arises here becomes decidedly marked with respect to conjunction. Evidentiation works in such a way that *conjunction is (evidentially) degrading*: the conjunction of two beliefs is in general less well evidentiated than either one of them. But with systemic centrality *conjunction is status-preservative*: the status of any *homogeneous* conjunction (conjuncts of items of the same status) is going to have the same as that of the conjuncts.[5]

Proceeding in this light, consider the following example, where our beliefs are as follows:

(1)   That creature is an elephant [classification: priority level 4].

(2)   Elephants are gray (not pink) [law: priority level 2].

(3)   That creature is gray (not pink) [description: priority level 4].

(4)   Only pigs and flamingos are pink [law: priority level 2].

And let us consider as a hypothesis not-(2): suppose that elephants (also) were pink.

Here there are two ways of restoring consistency:

(I)   Drop (1) and (4) and retain (2) and (3). Conditional: If elephants were pink, then that creature would not be an elephant.

(II)   Drop (3) and (4) and retain (1) and (2). Conditional: If elephants were pink, then that creature would be pink.

In effect we are led to a choice between (1) and (3). So with (3) as the weakest link in the order of precedence specified in display 1, we would be lead to alternative (II). We are thus led to endorse the counterfactual: "If elephants were pink, then that creature would be pink."

But now consider the situation of a variant supposition, namely the hypothesis

Not-(3): Suppose that that creature were pink (not gray).

As ever, different alternative routes to consistency restoration are available here:

(I)   Drop (4) and (2) and retain (1). Conditional: If that creature were pink, then there would be pink elephants.

(II)   Drop (1) and retain (4) and (2). Conditional: If that creature were pink, then it would not be an elephant (but a pig or flamingo).

In effect we are led to a choice between (1), and the conjunction of (2) and (4), the priority level of these three theses being 4, 2, and 2, respectively. Accordingly, with (1) here being in the weakest position in point of precedence we would be led to alternative (II). And so with (I) eliminated we arrive at the counterfactual:

If that creature were pink, then it would not be an elephant.

Again, consider the following situation:

$$X, Y$$

Let it be that the following facts are here seen as salient:

(1)   The box contains X.

(2)   The box contains Y.

(3)   The box contains several different letters.

(4)   The box contains only X and Y.

And now let it be that the question before us is: What would be in the box if (1) and (2) were false? Two possibilities will then arise:

(I)   We leave (3) intact, and now abandoning (4) arrive at: "If the box did not contain X and Y, then it would (still) contain some other letters in their place."

(II)   We leave (4) intact and now abandoning (3) arrive at: "If the box did not contain X and Y, then it would be empty."

Without any guidance as to the prioritization of (3) versus (4) we can do no more than arrive at the uninformatively disjunctive result:

If the box did not contain X and Y then either it would be empty or it would contain some other letters.

Getting us beyond this point would require the supposition-poser to be more specific regarding just what the issue is that he intends to put before us. He could (for example) reformulate his supposition as

Suppose that the box, *while continuing to be nonempty*, did not contain X and Y.

Then of course, alternative (II) would be precluded and (I) would prevail. The point is that in the absence of a fuller articulation of that determinative supposition, theses (3) and (4) have the same standing in point of MELF prioritization.

5   Natural versus Unnatural Counterfactuals

The different ways of resolving the problem of consistency restoration in the face of a counterfactual assumption may be entitled *scenarios*. These will clearly be different in point of plausibility, and a counterfactual will be plausible to whatever extent its conclusion obtains in the most plausible of the scenarios engendered by its antecedent supposition. Moreover, we can choose to transform this difference in degree in scenario plausibility into a difference in kind, by dichotomously separating those scenarios in the plausible and the implausible. And on this basis it will transpire that every conclusion that obtains in *every* plausible scenario engendered by a counterfactual supposition will yield a counterfactual that is categorically appropriate and tenable. (One would do well, however, to characterize it as such, rather than as *true*.)

The distinction between the appropriate and "natural" and the inappropriate or "unnatural" counterfactuals is, of course, going to play a crucial role. To illustrate this, let us suppose that we know that all the coins in the till are made of copper. Then we can say without hesitation:

If the coin I have in mind is in the till, then it is made of copper.

But we certainly cannot say counterfactually

If the coin I have in mind were in the till, then it would be made of copper.

After all, I could perfectly well have a certain silver coin in mind, which would certainly not change its composition by being placed in the till.

But just how is the difference between the two cases to be explained? Let $C = \{c_1, c_2, \ldots, c_n\}$ be the set of coins in the till, where by hypothesis all of these c_i are made of copper. And now consider the assumption:

Let x be one of the c_i (that is, let it be some otherwise unspecified one of those coins currently in the till).

Clearly this assumption, together with our given "All of the c_i are made of copper," will entail "x is made of copper" so that first conditional is validated.

But in the second case we merely have the assumption

Let x be a coin in the till (though not necessarily one of those currently there).

Now, of course, this hypothesis, joined to "All of the coins currently in the till are made of copper," will obviously not yield that conclusion. Accordingly, the second counterfactual is in trouble, since the information available to serve as its enthymematic basis is insufficient to validate the requisite deduction. The two conditionals are different because they involve different assumptions of varying epistemic status, a difference subtly marked by use of the indicative in the first case and the subjunctive in the second. For in the former we are dealing merely with descriptively *de facto* arrangements, whereas in the latter case we deal with a lawful generalization. And so generality prioritization speaks for the latter alternative. Lawfulness makes all the difference here.

Again, think of the virtually trivial counterfactual: "If I had struck the M key of my typewriter instead of the K key, it would have printed an M (and not a K)." Here I have the following beliefs:

(1) I struck the K key.

(2) I did not strike the M key.

(3) My typewriter printed a K (not an M).

(4) For any letter whatever, my typewriter prints out the letter that is struck on the keyboard.

The supposition is:

Not-(3): Suppose I had struck the M key.

Clearly (1) and (2) must be abandoned here. We are led to a choice between two alternatives: to abandon (4) in favor of retaining (3) or the reverse. Abandoning a general rule is undeniably more radical than abandoning a particular occurrence. And so we are led to the counterfactual:

If I had struck the M key, my typewriter would have printed an M and not a K

in place of the clearly more problematic counterfactual:

Even if I had struck the M key, my typewriter could (still) have printed a K.

Only if my auxiliary beliefs included some beliefs about my typewriter's systemic flaws or malfunctions would that second counterfactual obtain.

Consider the question "What if Booth had not murdered Lincoln?" And let us suppose that the salient beliefs here stand as follows:

(1) Lincoln was murdered in April 1865.

(2) Murder is deliberate killing, so if Lincoln was murdered, it was by someone deliberately trying to kill him.

(3) Booth murdered Lincoln.

(4) Only Booth was deliberately trying to kill Lincoln in April 1865.

Observe that (1), (2), (4) ⊢ (3); and now suppose that not-(3). Then we must abandon one of the trio: (1), (2), (4). Now (2) is a definitional truth, and (4) is a general fact, while (1) is but a matter of specific fact. So now the rule of precedence for matters of generality/informativeness marks (1) as the weakest link, and we arrive at:

If Booth had not murdered Lincoln, Lincoln would not have been murdered in April 1865.

In a similar vein, we have the problem of explaining how it is that the subjunctively articulated counterfactual

If Oswald had not shot Kennedy, then nobody would have

seems perfectly acceptable, while the corresponding indicative conditional

If Oswald did not shoot Kennedy, then no one did

seems deeply problematic.[6] Within the currently contemplated frame of reference the answer is straightforward. The background of accepted belief here is as follows:

(1)   Kennedy was shot.

(2)   Oswald shot Kennedy.

(3)   Oswald acted alone: no one apart from Oswald was trying to shoot Kennedy.

Now suppose that (2) is replaced by its negation not-(2), that is, suppose that Oswald had not shot Kennedy. For the sake of consistency we are then required to abandon either (1) or (3).

Our systematicity-geared policy of generality precedence now rules in favor of retaining (3), and thus dropping (1) and arriving at the former of that pair of conditionals. The alternative but inappropriate step of dismissing (1), would, by contrast, issue in the second, correspondingly implausible counterfactual. To be sure, this conditional could in theory be recast in a more complex form that would rescue it, as it were, to:

If Oswald did not shoot Kennedy then no one did, so since Kennedy was shot, Oswald did it.

In this revised version the conditional in effect constitutes a *reductio ad absurdum* of the idea that Oswald did not shoot Kennedy. But it is now clear that these conditionals address very different questions, namely the (1)-rejecting

What if Oswald had not shot Kennedy?

and the (1)-retaining

Who shot Kennedy?

respectively.

Again, let it be that we have as beliefs the following set of saliently accepted propositions:

(1)   Beggars are unable to realize their wish for horses [specific fact about beggars].

(2)   People who have an effective mode of transport at their disposal (e.g., who have horses) will generally make use of it (i.e., will ride those horses) on appropriate occasions [general fact about a human modus operandi].

(3)   Beggars do not (generally) ride horses [comparatively specific fact about beggars].

And now let us project the (belief-incompatible) supposition not-(1), by supposing that beggars *could* realize their wishes for horses. Clearly when not-(1) is adjoined to (2) and (3) we obtain a collectively incompatible set of commitments, so that either (2) or (3) must be abandoned. And given the policy of generality-preference, it is clear that (3) must give way to (2), and we thus validate the counterfactual:

If wishes were horses, then beggars would ride.

Assume that someone has placed a bird (we know not of what species) into a box. Since we know that all ravens are black we unproblematically have it that

If that bird is a raven, it will be black.

However, even if we knew that the bird in the box to be a cardinal, we still have:

If that bird were a raven, it would be black.

We obtain this because we know

(1)   The bird in the box is not a raven.

(2)   All ravens are black.

(3)   The bird in the box is not black.

And now when we look to the antecedent of that conditional we assume that not-(1). We must then reject either (2) or (3). And in view of generality preference it is (3) that must go. And we thus obtain the above-indicated conditional instead of the inappropriate

If that bird in the box were a raven then it would not hold that all ravens are black (since that bird is not black).

However, consider replacing (3) among the above premises with the equally true

(3′)   All birds in the box are non-black.

Within the ground (1′), (2), and (3′) we would now be confronted with a choice between retaining (2) and (3′). It might now seem that generality preference does not help at this point, since both (2) and (3′) are truths of universal format (All Xs are Ys). But this consideration is general because the scope of (2) is greater, since (2) is spatiotemporally unrestricted while (3′) is box delimited. As a law of nature (2) holds in all physically realizable situations, something that certainly cannot be said for (3′).[7]

Again consider the "Carolina Problem" proposed by Nelson Goodman.[8] Supposing Jones to be in Georgia, we have the following beliefs:

(1)   Jones is not in South Carolina.

(2)   Jones is not in North Carolina.

(3)   North and South Carolina comprise the Carolinas.

Given (3) we are bound to have:

(4)   If Jones were in the Carolinas, he would be either in North Carolina or in South Carolina.

To see how this counterfactual is validated, let us now project the counterfactual hypothesis:

(5)   Suppose that Jones is in the Carolinas.

Here we have it that

(1), (2), (3), (5) ⊢ (contradiction).

In the interests of consistency, one of those four theses must clearly be abandoned. But (5) holds by stipulation and (3) enjoys generality preference. So, seeing that (1) and (2) stand on exactly the same level, we arrive at

If Jones were in the Carolinas, he would be either in North Carolina or in South Carolina (we can't say which).

Thus, counterfactual (4) emerges as appropriate.

However, we will have (4) without having either

(6)   If Jones were in the Carolinas he would be in North Carolina

or

(7)   If Jones were in the Carolina he would be in South Carolina.

And this is only to be expected. After all, in the factual case one can unproblematically establish

If the coin is tossed, it will come up heads or tails

without being able to establish either

If the coin is tossed it will come up heads

or

If the coin is tossed it will come up tails.

Again, consider the counterfactual "If Puerto Rico were a state of the Union, the U.S. would have 51 states." Here the following three propositions may be taken as salient given beliefs:

(1)   Puerto Rico is not a state of the Union.

(2)   There are 50 states.

(3)   The list of the states includes *all* the following: Alabama, Arizona, etc.

(4)   The list of the states includes *only* the following: Alabama, Arizona, etc.

Now consider introducing the fact-contravening supposition: "Suppose that not-(1), that is, suppose Puerto Rico to be a state." In embarking on this assumption we obviously now have to jettison (1) and (4) and replace them by their negations. But this is clearly not enough. The set {not-(1),

(2), (3)} still constitutes an inconsistent trio. To keep within the limits of consistency we must also abandon either (2) or (3), and so we can either retain (2) and loose some existing state or other, or we can abandon (2) and increase the number of states by one. Of course we have a choice between these two alternatives. And with it we arrive at a choice between two alternative counterfactuals.

(A)   If Puerto Rico were a state, then there would be 51 states since all of the following are states: Alabama, Arizona, etc. (Here we retain (3) and abandon (2).)

(B)   If Puerto Rico were a state, then one of the present states would have departed, since there are just 50 states. (Here we retain (2) and abandon (3).)

As usual, the question comes down to a matter of priority and precedence. And here (3) prevails on grounds of systemic informativeness. The first of the indicated counterfactuals is accordingly more eligible then the second.

Nelson Goodman has posed the problem of the cotenability of such conflicting conditionals as

If Napoleon were Julius Caesar, then Napoleon would be dead by 100 CE (since Caesar was).

If Napoleon were Julius Caesar, then Caesar would be alive in 1800 CE (since Napoleon was).

We here have a situation of incompatibility: the two conditionals, while seemingly innocuous when taken separately, are nevertheless not cotenable. One might be tempted to take both of these in stride and see the pair as basis for a *reductio ad absurdum* of the "Napoleon = Caesar" hypothesis. But flatly rejecting a hypotheses is never a congenial option. So let us look at the matter in a different light.

The issue arises in a context where the salient beliefs are:

(1)   Caesar was dead by 100 CE.

(2)   Napoleon was alive in 1800 CE.

(3)   If $X = Y$, then whatever holds of X will hold of Y.

Let us now project the hypothesis:

(4)   Napoleon = Caesar.

We then have it that:

(1), (3), (4) ⊢ ∼(2)

(2), (3), (4), ⊢ ∼(1)

Given (4), at least one of (1), (2), or (3) must be abandoned. Since generality precedence rules in favor of retaining (3), and moreover (1) and (2) stand on the same plane, the most and the best we can to is to validate the conditional:

If Napoleon were Julius Caesar, then either Napoleon would be dead by 100 CE or Caesar would be alive in 1800 CE.

The two abovementioned counterfactuals that constitute an inconsistently conflicting pair simply cannot be validated. That "Cotenability Problem" vanishes thanks to the absence of cotenability.

Again, consider the following counterfactual example, due to David Lewis:[9]

If kangaroos had no tails, they would topple over.

This occurs in a context where we believe:

(1)   Kangaroos have tails.

(2)   Kangaroos are stable and do not topple over.

(3)   The S-shaped stance of kangaroos involves a raised head-and-forebody.

(4)   The principle of the counterlever requires that maintaining a head-and-forebody in such a raised position must be balanced by a counterweight.

(5)   Given the structure of kangaroos, the requisite counterlevering counterweight is provided—and only provided—by the tail.

Let us now project the supposition of not-(1), that is, that kangaroos have no tails.

In conjunction with (3) through (5) this assumption of not-(1) yields the conclusion that kangaroos will become unstable and so topple over, thus conflicting with (2). Thus one of (2) through (5) must be abandoned. Here too we have to choose between various alternatives:

(3), (4), (5)/(1), (2)

(2), (4), (5)/(1), (3)

(2), (3), (5)/(1), (4)

(2), (3), (4)/(1), (5)

But since their shape is a characterizing feature of kangaroos as such as per (3), and (4) and (5) are general principles, the case-specific nature of (2) makes it the weakest link, whose sacrifice in the circumstances validates the counterfactual at issue. (And of course if "deeper" facts about evolution were added to that list, (2) would become even more vulnerable.) Of course it is not that (2) and (3) through (5) are incompatible as such, flat out and absolutely. But they become so in the presence of the (clearly substantive) supposition of not-(1).

All in all, then, systemic fundamentality as per the MELF priority ordering is a crucial factor in analyzing counterfactuals.

6 Complications

Some further complications yet remain. An instructive example is afforded by the counterfactual:

If this cube of lead were made of sugar, then it would be water-soluble.

Our salient beliefs are:

(1)   This cube is made of lead.

(2)   This cube is not made of sugar.

(3)   Lead is not water-soluble.

(4)   Sugar is water-soluble.

And let us suppose not-(2): This cube is made of sugar.

We must obviously abandon (1) in the wake of this supposition. And the salient beliefs are now few enough to be consistent, so that the conclusion of that counterfactual follows straightaway from our supposition and (4). Here—somewhat unusually—no further recourse to comparative prioritization is required.

However, one could complicate the story by switching to the more complexly formulated counterpart counterfactual:

If this cube of lead were made of sugar, and if it were placed in a body of (ordinary) water for a long time, then it would dissolve.

Our relevant beliefs are:

(1)   This cube is made of lead.

(2)   This cube is not made of sugar.

(3)   Lead is not water-soluble.

(4)   Sugar is water-soluble.

(5)   This cube has not been placed in water for a long time.

(6)   This cube did not dissolve.

And now we suppose:

(S)   Not-(2) and not-(5).

We must now, of course, reject not only (2) but also (1) and (5). But thereupon we can move to not-(6) from

Not-(2), (4), and not-(5).

And this validates the counterfactual at issue, given that the alternative, namely rejecting (4), would violate the prioritization rule of generality precedence. So now enlarging the scope of saliency restores our usual dependency on such a rule.

And there are further complications. Thus consider the conditional

If that pencil conducted electricity, then (since it is made of wood) wood would conduct electricity.

Here the salient beliefs are:

(1)   Wood does not conduct electricity.

(2)   That pencil is made of wood.

(3)   That pencil does not conduct electricity.

These are related so that: (1), (2) ⊢ (3). Accordingly, the supposition of not-(3), serving as antecedent of the conditional, forces the abandonment of (1) or (2) (or possibly both). Ordinarily we would, under the aegis of law-prioritization, make (2) give way to (1) and arrive at:

If that pencil conducted electricity, then it would not be made of wood.

However, the parenthetical stipulation of the initial conditional explicitly prioritizes (2)-retention. And in view of this we must now make (1) give way to (2) so as to arrive at the initial counterfactual as stated.

The crux here is that whoever poses the counterfactual question at issue—"What if that pencil conducted electricity?"—will, thanks to questioner's prerogative, have both the privilege and the obligation to define the terms of reference under which the question is posed. And while the normal presumption sees it as merely speculative so that the second conditional would be the natural resolution there remains the option of so representing the issue that it is the first conditional that is appropriate.

This issue of questioner's prerogative becomes particularly relevant in the setting of the distinction between factual inquiry through experimentation and purely hypothetical speculation. This difference is best clarified by means of an illustrative example.

Let it be that a chemist considers the thesis:

(1)   Acids turn blue litmus paper red.

And let it also be that she tests this thesis by an experiment as follows:

(2)   Beaker B contains some acid.

(3)   Slip S consists of blue litmus paper.

(4)   Slip S is immersed in beaker B.

Here (1) through (4) yield the circumstance

(5)   Slip S turns red.

But suppose not-(5). Then in the circumstances at issue it will transpire that we would arrive at

If that slip of litmus paper did not turn red, then (1) would be false.

In an experimental context of this sort, it is the theory being tested that is at risk: (1) is the weakest link.

But consider a different situation: one in which (1) through (3) are all taken as accepted facts along with two further items, namely

Not-(4): The slip is not immersed in the acid.

Not-(5): The slip does not turn red.

And let us now adopt by way of stipulation the counterfactual hypothesis

(4)   Slip S is immersed in beaker B.

We would now of course arrive at:

If that slip had not turned red upon being immersed in beaker B, then either it would not be litmus paper or the beaker would not contain acid.

Here, in the face of that counterfactual postulation, it is (as usual) generality-preference that comes into operation so that (1) is safe and (2) through (3) are at risk. The important lesson here is that counterfactuals are not of a piece. They can be *speculative* in the manner of the thought experiment of that central situation or they can be *stipulative* in the manner of the second situation.

One further matter. The question, "What if that slip did not turn red?" can be used in two different ways: as the thought-experimental test for a generalization which puts that generalization at risk, or as a stipulative hypothesis that puts certain factual circumstances at risk. Of course, which of these alternatives is in view is a matter of questioner's prerogative—a matter of the issue-poser's decision as to exactly what sort of question is to be at issue. And this will make a great deal of difference. For in inductive/evidential situations we subordinate theories to facts, but in full-fledgedly counterfactual contexts, where our suppositions are purely hypothetical the order of prioritization is reversed.

In discussing the present author's earlier treatment of some of these issues, one critic complained about there being no *justifactory rationale* for the prioritization of generality and lawfulness other than its retrospective efficacy of yielding the right results regarding counterfactuals.[10] But this charge is doubtful in relation to that earlier account[11] and clearly wrong as regards the present one. For breaking that chain of inconsistency at its systematically weakest link is clearly a matter of damage control, of sacrificing as little informatively useful material as possible. And in fact this is just what that earlier treatment maintained: "We class the relevant beliefs into distinct categories so as to indicate how fundamental they are in the scheme of our knowledge, and then let such an ordering be our guide in effecting the reconciliation demanded be a belief-contravening hypothesis."[12] How much clearer can one make it that the governing rationale here is the desideratum of saving as much as we can by way of systematically more serviceable beliefs?

There are clearly two ways to arrive at a belief-contravening thesis. The first is the *inductive* route by way of a new discovery, the acquisition of a new fact inconsistent with our established belief system B. The second is the *hypothetical* route by way of a mere supposition and assumption. In both cases the imperative of consistency compels belief-revision: in both cases we must revamp our belief system B to come to terms with that new item. Now when new discoveries are at issue we prioritize definite facts over our generalized commitments and make theories give way to well-established facts. But when merely stipulative hypotheses and assumptions are at issue, we engage in generality preference and prioritize general theories over particular facts in the competition for survival. It is the factor of overall informativeness that now becomes paramount.

7 Factual Contexts Involve a Different Perspective

The story is told that Herbert Spencer said of Thomas Buckle (or was it the other way round—as it could just as well have been) that his idea of a tragedy was a beautiful theory destroyed by a recalcitrant fact. A fundamental epistemic principle is at issue here. Among seemingly plausible propositions in factual contexts, when a limited particularity and a broad generality come into conflict, it is the former that will prevail. Facts, as the proverb has it, are stubborn things. In the case of a clash, facts prevail over theories, observations over speculations, concrete instances over abstract generalities, limited laws over broader theories. Specificity predominates generality in factual contexts. Here the cognitive dissonance that needs to be removed must be resolved in favor or the more particularly concrete, definite party to the conflict. Generality is a source of vulnerability: and when clashes arise, particularity bespeaks primacy. For the more general, the more cases included, and so the more vulnerable. Accordingly, in "normal" circumstances specificities are on safer ground. We have to presume that specifics are in better probative condition than generalities because they are by nature easier to evidentiate. Specifics are easier to establish than generalities because generalities encompass a multitude of specifics. Counterwise, once established generalities are easier to disestablish than specifics because a single counterinstances among many possibilities will disestablish a generality, whereas it takes certain particular cases to disestablish a particularity.

The more broadly informative, the greater the range of generality (as opposed to limitedness), the higher the detail and specificity (as apposed to vagueness) the more vulnerable, the less secure a claim becomes. A "complementarity relationship" of sorts obtains. (See figure 9.1.)

In any case, what we are dealing with here—what is basically at issue with specificity preference—is not a general truth-claim but a procedural principle. What is at issue is not a proposition but a precept of epistemic practice such as "Believe the testimony of your own eyes" or "Accept the claim for which the available evidence is stronger." Of course we can go wrong here: It is not true that what your eyes tell is always so or that the

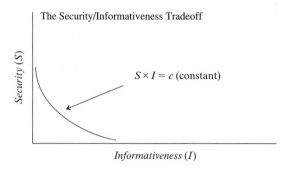

Figure 9.1
The overall quality of the information provided by a claim hinges on combining its security with its definiteness. Given suitable ways of measuring security (S) and informativeness (I), the curve can be supposed to be an equilateral hyperbola obtained with $S \times I$ as constant.

truth always lies on the side of the stronger evidential case in hand. But while not infallible they are good guides to practice. And the justification of such a principle of practice is ultimately a matter of general adequacy rather than fail-proof correctness. The justification at issue is thus one of pragmatic efficacy—of serving the purposes of the practice at issue effectively. It is a matter less of demonstrating a universal truth than of validating a modus operandi on the basis of its general efficacy.

The principle of specificity precedence in factual contexts can be illustrated from many different points of view. As already noted, it is a standard feature of scientific practice: when theory and observation clash, it is observation that prevails.[13] Whenever speculation clashes with the phenomena, a hypothesis with the data at our disposal, or a theory with observation, it is generally—and almost automatically—the former that is made to give way.

And an analogous situation prevails at a higher level as well. What is now in view is not the clash of observation with theory, but rather the clash of a lower level (less general or abstract) theory with one of a higher (more general and abstract) level.

Moreover, this situation for the experimental sciences finds a class analogue in the case of the historical sciences. A single piece of new textual evidence or a single item of new archeological discovery can suffice to call a conflicting theory into question. A penchant for specificity preference is very much in operation throughout the range of factual inquiry.

A further illustration of specificity preference emerges in the context of philosophy. The work of Thomas Reid (1710–1796) and the philosophers of Scottish school manifest this in an especially vivid way. These thinkers reasoned as follows: Suppose that a conflict arises between some speculative fact of philosophical theorizing and certain more particular, down-to-earth, bits of everyday common sense. Then it of course will and must be those philosophical contentions that must give way.

Reid insisted that common sense must hold priority over the more speculative teaching of philosophy. Maintaining that most philosophers themselves have some sense of this he observes wrongly that "it is pleasant to observe the fruitless pains which Bishop Berkeley took to show that his system ... did not contradict the sentiment of the vulgar, but only those of the philosophers."[14] Reid firmly held that any clash between philosophy and common sense must be resolved in the latter's favor. Should such a clash occur:

The philosopher himself must yield ... [because] such [common-sense] principle are older, and of more authority, than philosophy; she rests upon them as her basis, not they upon her. If she could overturn them she would become buried in their ruins, but all the engines of philosophical subtlety are too weak for this purpose.[15]

In any conflict between philosophy and everyday common-sense beliefs it is the latter that must prevail. The lessons of ordinary experience must always prevail over any conflicting speculations of philosophical theorizing. On this point the Scottish common-sensists were emphatic. When conflicts arise, commonplace experience trumps philosophical speculation. And in fact the bulk of metaphilosophical accounts of philosophizing have agreed with this specificity-forming point of view.

Yet another illustration of specificity preference in inquiry comes (perhaps surprisingly) from pure mathematics. In deliberating about the relationship between mathematics proper and metamathematical theorizing about mathematical issues, the great German mathematician David Hilbert (1862–1943) also argued for specificity preference. If any conflict should arise between substantive mathematical findings and metamathematical theory, so he maintained, it is automatically the latter that must yield by way of abandonment or modification. Metamathematical theses are the more vulnerable because they always reach across a wider range of mathematical fact. Accordingly, consider what Arthur Fine calls "Hilbert's Maxim," namely, the thesis that

Metatheoretic arguments [about a theory] must satisfy more stringent requirements [of acceptability] than those placed on the arguments used by the theory in question.[16]

This too affords yet another illustration of specificity preference.

And so, in all such cases a clash of contentions is resolved by having the more general party to the conflict give way to the more definite by preferring specificity to abstractness. Throughout it would seem our pursuit of knowledge prioritizes specificity. Rational inquiry is governed by a tropism to specificity.

Is there a cogent rationale for this prioritization of particularity? Are there sound reasons of general principle why specificity should be advantaged in matters of inquiry?

The answer is clearly an affirmative one. The reasoning at issue runs somewhat as follows. Consider a conflict of the sort that now concerns us. Here in the presence of various other uncontested "innocent bystanders" (x) we are forced to a choice between a generality (g) and a specificity (s) thanks to the logic of the schematism inherent in the (equivalent) conflict-indicating implications:

$(g \,\&\, x) \rightarrow \sim s$, or equivalently $(s \,\&\, x) \rightarrow \sim g$

With x fixed in place, either s or g must be sacrificed. But since g, being general, encompasses a whole variety of other special cases—some of which might well also go wrong—we have, in effect, a clash between a many-case manifold and a fewer-case situation. Since the extensiveness of the former affords a greater scope for error, the latter is bound to be the safer bet. Generalities are more vulnerable than specificities. When other things are anything like equal, it is clearly easier for error to gain entry into a larger than into a smaller manifold of claims.

To be sure, this line of thought rests on a presumption of other-things-equal that may fail to be realized. It is certainly conceivable that all is well with that g-generalization and that something has gone wrong with our s-specificity; or, that other-things-equal presupposition x may be off the mark so that with both s and g as alike true (or alike false) it is x that is the cause of trouble. But barring such anomalies, specificity can reasonably be expected to prevail.

8   A Crucial Inversion

It is striking that the standard presumption at issue with specificity prioritization is in fact inverted and the reverse procedure of generality prioritization obtains when we turn from factual to counterfactual contexts.

In factual contexts we prioritize evidentiation; here in cases of conflict the more strongly evidentiated proposition wins out. But in counterfactual contexts we proceed differently. Here it is fundamentality rather than evidentiation that matters.

This crucial difference between factual and counterfactual situations in relation to evidentiation can be illustrated as follows. Consider the following situation. Our salient beliefs are:

(1)   John married Mary [strongly evidentiated belief].

(2)   John married Jane [weakly evidentiated belief].

(3)   John is a bigamist [firm belief].

Now in the factual situation at hand let us suppose we were to contemplate its turning out that in fact:

Not-(3): John is no bigamist.

We would arrive at the factual conditional

Since John is not a bigamist, then (almost certainly) he did not marry Jane. If not-(3) then not-(2).

In such factual cases, the chain of inconsistency has its weak link at the evidentially weakest point.

However in the counterfactual (purely speculative) cases we do not and should not reason in this way. Here we would *not* conclude

If John were not a bigamist, then he would (very likely) not have married Jane.

All we can arrive at is

If John were no bigamist, then he would certainly not have married both Mary and Jane.

In this counterfactual setting those factual theses (2) and (3) remain on a par in our reasoning; their sameness of systemic status (in the sense of MELF) predominates over their difference in point of mere evidentiation.

In the context of counterfactuality, rational procedure becomes a matter of keeping our systemic grip on the manifold of relevant information as best we can. Once we enter the realm of fact-contravening hypotheses, those general theses and themes that we subordinate to specifics in factual matters now become our life-preservers. We cling to them for dear life, as it were, and do all that is necessary to keep them in place. "Salvage as much information about the actual state of things as you possibly can" is now our watchword. Accordingly, specifics and particularities will here understandably yield way to generalizations and abstractions.

This line of thought reinforces a point for which the present author has argued for many years, namely that lawfulness (in the sense of natural law) and generality of range are pivotal features in the treatment of counterfactuals.[17] Once we enter the realm of fact-contravening hypotheses and suppositions, those general truth and theories become our sheet-armor. We cling to them for dear life, as it were, and do all that is necessary to keep them in place. Almost universally, specifics and particularities yield way to generalizations and abstractions. The overall lesson then is clear. When a clash among seemingly acceptable propositions occurs in *factual* contexts, considerations of evidential plausibility lead us to adopt the stance of specificity-preference. But in counterfactual contexts where the economics of information management is paramount, our deliberations must pivot the generality preference at issue with systemic cogency.

To be sure, in the case of a counterfactual supposition that is itself particular we may have to make a generalization give way to it. This arises standardly in the case of thought experiments that contemplate outcomes that may defeat generalizations. Thus consider the following counterfactual relating to the generalization (g) that heavy objects (like rocks) fall to earth when released:

If this heavy rock had not fallen to earth when it was released at altitude yesterday, then g would be false.

Here we have the following beliefs regarding the facts of the situation:

(1)   That heavy rock was released at altitude yesterday [fact].

(2)   That rock then fell to earth [fact].

(3)   Heavy objects (like rocks) fall to earth when released at altitude (g) [law].

When now instructed to assume not-(2), the resulting inconsistency forces a choice between abandoning the specific (1) and the general (3). With generality precedence at work we would be led to retain (3). But that of course is not how things work in such a thought experiment. For now the particular thesis at issue—namely (1)—is immunized against rejection by the circumstance of its constituting part of the very hypothesis under consideration.

9 Summary

Counterfactuals are simply a special case of apory resolution required by the introduction of belief-contravening suppositions. And the rational approach here is that of making the optimal—minimally disruptive—readjustment in our accepted beliefs compatible with introducing that belief-contravening assumption among them.

However, the question "Where is the weakest link in a chain of interrelated beliefs?"—"Which is the weakest thread in a web of belief?" is ambiguous and can be construed in two importantly different senses, namely weakest in point of *evidentiation* on the one hand, and on the other the weakest in point of what might be called *systemic centrality* or informativeness.[18]

As regards evidentiation every belief stands on its own—the question is simply: how well does the available information substantiate it in any specific case? But as regards systemic centrality, the question is one of the *types* of contention at issue and relates not to the specific information that it conveys but to the generic mode and manner in which it conveys information. It is a matter not of the specific and particular content of a belief but of the generic informativeness of the sort of contention at issue. Here we proceed not inductively but categorially.

What is determinative in the context of counterfactuality is the function of systemic enmeshment in the manifold of knowledge. Intuitive appeal, evidential support, judgmental plausibility are not at issue. For the ruling principle here is not psychological (inductive appeal,[19] psychological alteration[20]), and evidential considerations (particularly, intuitive support, etc.) have nothing to do with it. Moreover, what suppositional reasoning, properly understood, involves is emphatically not a matter of an intuitive insight into the modus operandi of nonexistent possible worlds.[21] It is, rather, a matter of bringing the standard principle of information management to bear upon our experimentally grounded knowledge of matters of actual fact. The priority among conflicting propositions in aporetic settings is, accordingly, not a matter of the personal preferences and predilections of individuals; it is determined objectively relative to the purposive orientation of different contexts of deliberation.[22]

Consider an example. It is a known fact that:

(F1)   People can be identified via their fingerprints: Only X can have X's fingerprints.

But it is also a known fact that:

(*F*2)   Spatially separated individuals are distinct people: *X* can occupy only *X*'s position, other positions have to be occupied by others: someone cannot be in two different places at once.

We could try to play these facts off against each other by introducing the following supposition:

(*S*)   *Y*—who occupies a spatial position different from *X*'s—also had *X*'s exact fingerprints.

(*S*), (*F*1), and (*F*2) form an inconsistent trend. In the face of our supposition we are construed to give up (at least) one of those two thoroughly well-established generalizations. What to do?

Here generality precedence will not help us—it is between generalities that we must choose. However (*F*1) is merely an empirical generalization that reflects a contingent fact about the world's makeup that we have learned to be true through empirical investigation. On the other hand (*F*2) is a conceptual fact built into a fundamental commitment to the idea of a physical object with its inherent principle that—such an object cannot be in two different places at one and the same time. Thus we give precedence to (*F*2) or (*F*1)—not because it is better evidentiated by the empirical facts, or because we inevitably find that we have a greater belief inclination toward it, but rather because the fact it represents plays a systemically more fundamental role in the cognitive scheme of things.

The crux of counterfactual analysis is thus *not a matter of scrutinizing the situation of other possible worlds but rather of categorically prioritizing our beliefs regarding this actual one.* For most fortunately, the analysis of counterfactuals does not require recourse to anything as grandiose as "possible worlds." All that is needed is a comparative prioritization of the relevant beliefs that concern a mere handful of propositions that figure immediately in the particular case at hand.

The process of resolving aporetic conflicts rests on a master principle that runs essentially as follows:

In restoring consistency maintain overall credibility. Give priority to those propositions that provide for the systematically optimal *combination* of plausibility (evidentiation) *and* problem resolving (informativeness). Treat the demands of smooth systematization as paramount. [The process here is thus a hybrid mix—a balanced *combination* or *fusion* of the evidential and the hypothetical approaches.]

Accordingly, in factual matters we must take evidential considerations into account; in purely counterfactual contexts we do not worry ourselves about them. Here systemic fit and cognitive fundamentality rule the roost.

The difference between the specificity prioritization in factual contexts and the generality prioritization that is operative in hypothetical ones serves to illustrate the following important points:

1. The principles of counterfactual reasoning are not matters of abstract theory but rather of information-processing policy.

2. The policies at issue are not determined by considerations of theoretical logic but find their rational validation in considerations of utility and efficacy within the purposive contexts at issue.

3. For this reason, counterfactivity should—strictly speaking—not be classed as true/false but rather as appropriate/inappropriate, warranted/unwarranted, acceptable/unacceptable. Not truth but rational cogency is the operativetive factor here.

Overall, the paramount lesson is that the mode of procedure in the different areas of reasoning is governed by the context-specific aims and objectives at issue. Here, as elsewhere, the pragmatic factor of purposive efficacy becomes the determinative consideration.

Postscript: Historical Observations

The thesis that in specifically counterfactual situations a principle of generality preference obtains was urged by the present author originally in a 1961 paper[23] and developed more fully in my *Hypothetical Reasoning* (Amsterdam: North Holland, 1964). Subsequently, a series of empirical investigations by R. Revlis and his colleagues confirmed experimentally that in actual practice people do indeed proceed in this way.[24] And among philosophers too the idea of prioritizing lawful generality has become widely accepted as a matter of empirical practice grounded in psychological inclinations.[25] Such psychologism is, however, decidedly different from my own position, which sees the prioritization of lawful generality as a matter of functional efficacy in the light of the inherent objectives of counterfactual communication. At bottom the construal of counterfactuals is not a psychological matter of preference or reluctance to change, but one of information-processing prioritization rooted in the function-oriented ground rules of linguistic practice. That people do in fact think in this way is the result of, rather than the ground for, the validity of the principle. It does no more than to reflect the happy circumstance that in this particular area people generally proceed in a rationally appropriate way.[26]

10 Further Complications of Counterfactuality

1 Problems of Information Shortfall

Consider the following situation, where the question at hand is "What if Napoleon had won at Waterloo?" Our beliefs are:

(1)   Napoleon lost the battle of Waterloo.

(2)   Napoleon attempted to flee France on an American ship a fortnight after the battle of Waterloo.

(3)   Napoleon was captured by the British and sent into imprisonment on St. Helena about a month after the battle of Waterloo.

(4)   Napoleon died in exile on St. Helena some six years after the battle of Waterloo.

And let us now make the following supposition:

Napoleon won the battle of Waterloo.

How are we to revise our beliefs in the face of this assumption? It is obvious that (1) must go—this is a simple matter of stipulation. But what about (2) and (3)? They plainly do not stand in any *logical* contradiction to the hypothesis we are asked to make. But in the wider cognitive setting of the question at hand these two theses accord so poorly with this supposition that it is clear that they too must be jettisoned. But what of (4)? Here we can say little; it is uncertain just what ought to be done about its retention or rejection. Here the situation is informatively too remote: in the circumstances we simply do not have enough information to resolve the issue. It is somewhere between unclear and doubtful that there are any considerations within the wider manifold of our relevant beliefs whose prioritization requires abandoning (4).

Thus suppositional questions sometimes arise in the setting of a dearth of information so grave as to preclude any counterfactual response whatever. We can certainly wonder "If Adolf Hitler had been killed in World War I, who would have been the German chancellor in 1940?" Yet for this question, as for so many others, no sustainable recourse can be secured. Our informative resources are inadequate to sustain any resolution here. There just are no tenable counterfactuals of the form:

If Adolf Hitler had been killed in World War I, then X would have been the German chancellor in 1940.

Again, let it be that a stone sits on the stand before us and that:

(1) This stone (S) weighs over three pounds (symbolically: $w(S) > 3$ lbs.).

And now suppose not-(1):

Not-(1) This stone weighs three pounds or less (symbolically: $w(S) \leq 3$ lbs.).

Clearly this supposition of not-(1) will of itself yield the consequence:

$(\exists x)(x \leq 3 \,\&\, w(S) = x)$

And so we unproblematically and trivially have:

$(w(S) \leq 3) \,\{B\}\!\!\rightarrow (\exists x)(x \leq 3 \,\&\, w(S) \leq x)$

or, verbally:

If the weight of the stone were less than three pounds (which it isn't), then there would be some particular weight, less than three pounds, which would be the weight of that stone.

But it is clearly not the case that

$(\exists x)([w(s) \leq 3] \,\{B\}\!\!\rightarrow [x \leq 3 \,\&\, w(s) \leq x])$

or, verbally:

There is some particular weight less than three pounds such that if the weight of the stone were less than three pounds, then the stone would be of that weight.

There is, in the circumstances, no way of validating this counterfactual contention. In point of the question: "So just what would the weight of the stone be if it weighed less than it does?" there just is no determinable fact of the matter. But nevertheless that first counterfactual stands secure.[1]

But why is there no cogent answer to this question? It is not because of unavailable worlds, but simply because of an informational insufficiency, an inability of the available information regarding the hypothetical circumstances to yield a result of the desired specificity. It is like asking "If that dog were not a collie, what would it be?" or "If that die toss had not come up 1, then how would it have come up?" There is no viable answer to such a question save the trivial: "Something different."

2 Seemingly Problematic Cases

The presently contemplated analysis of counterfactuals has it that the validation of such conditionals hinges on the background of our beliefs, so that fact typically takes precedence over counterfact. What we have here is an existentialism that prioritizes reality over possibility. Assertible counterfactuals rest on a factual basis: the actual facts being as they are is a necessary, *sine qua non* condition for the tenability of counterfactuals. To be sure, the facts, though needed, are not in themselves altogether sufficient to validate counterfactuals; a matter of evaluation—namely, the comparative prioritization of facts—will also be an indispensable part of the overall picture.

It is instructive to consider some further illustrations of such an epistemically based (rather than semantical) approach to the validation of counterfactuals. Take the counterfactual contention: "If the Eiffel Tower were in Manhattan, then it would be in New York State." This conditional introduces a fact-contravening hypothesis:

The Eiffel Tower is in Manhattan

into a belief context where acceptance of the following three statements is salient:

(1)   The Eiffel Tower is in Paris, France.

(2)   The Eiffel Tower is not in Manhattan.

(3)   Manhattan is in New York State.

With (2) replaced by its negation we have it that, as usual, the result of replacing a thesis by its counterfactual negation still leaves an obviously inconsistent situation. In seeking to remove the resulting inconsistency, we obtain the following acceptance/rejection alternatives:

(1)/(2), (3)

(3)/(1), (2)

We are thus confronted with the choice between (1) and (3). And of course the latter will have the upper hand in point of generality and scope through contracting larger-scale locales as per (3) with the location of specific structures as per (1). Accordingly thesis (1) would have to be abandoned as well and the second alternative schedule adopted. We would thus arrive at the conditional: "If the Eiffel Tower were in Manhattan, then it would not be in France but in New York State."

In theory, to be sure, our counterfactual hypothesis seems to leave both of those alternatives open, namely the more natural one just stated and its anomalous rival:

If the Eiffel Tower were in Manhattan then Manhattan Island would be in Paris (since that is where the Eiffel Tower is).

The two counterfactuals confront us with a choice between moving an individual structure and moving a whole island. And subject to the policy of the economy of information management inherent in systematicity precedence—of granting precedence to those contentions that are systemically more informative through being more general and far-reaching—we shall keep that island in place.

Again, let it be that the following three contentions qualify as known facts:

(1)   All whales are mammals.

(2)   This creature is a whale.

(3)   This creature is a mammal.

And let us assume:

Not-(3) = This creature is not a mammal.

This assumption, of course, constrains the rejection of (3). But even then, the residual group (1), (2), not-(3) is still inconsistent. To restore consistency in the face of our assumption we must accordingly drop either (1) or (2) and are confronted with two alternative counterfactuals:

(I) If this creature were not a mammal then it would not be the case that all whales are mammals (and so, some whales are not mammals), because this creature (which is a whale) is not a mammal. (Here we drop (1) in favor of (2).)

(II) If this creature were not a mammal then it would not be a whale, because all whales are mammals, and by hypothesis, this creature is not. (Here we drop (2) in favor of (1).)

The issue comes down to retaining (1) versus (2) in the face of not-(3). And here our generality preference will now automatically favor alternative (II) over (I).

This particular result may appear to be counterintuitive. But the reason for this is readily explained. The law-prioritization at issue in our analysis of counterfactuals is subject to one crucial proviso, namely that the generalization at issue is indeed substantively lawful. For since *informativeness* is the pivotal consideration here, mere *grammatical generality* will not do. (Otherwise we would prioritize such merely accidental generalizations as: "All coins in the box are pennies," to arrive at "If this dime were a coin in the box, it could be(come) a penny.") Now in the example it is, in theory, possible to regard "All whales are mammals" as an accidental generalization. For although scientific sophisticates may deem it a law of nature, "the man in the street" would perhaps take the following line:

Being a whale is a matter of the surface structure of a creature—of its outer appearance and general geometry. On the other hand, being a mammal is a matter of inner machinery—having a mammary gland that can supply milk for offspring. Accordingly it would be perfectly possible to find creatures that are whales to every outward appearance, but which nevertheless are not mammals. "All whales are mammals" is thus something incidental or accidental: mammalhood just is not a lawfully requisite feature of being a whale. And so when we assume "This [whale-configured] creature is not a mammal" we destabilize that generalization exactly as we destabilize "All coins in the box are pennies" when we make the supposition "Assume that this nickel were a coin in the box." Here generality precedence no longer applies.

On such a perspective, the seemingly nonconforming option of alternative (I) actually instantiates rather than invalidates the present treatment of counterfactuals. Generality preference is not refuted here—it is just that those involved do not regard "All whales are mammals" as being a lawful generalization.

Sometimes counterfactual information is crucial to other counterfactuals because counterfactuals can have other counterfactuals as parts of their epistemic basis. Thus consider the counterfactual:

Even if Mary had rejected his offer of marriage, John would (still) have become her father's son-in-law (because he would then have married her sister Jane).

Here some background belief along the lines of "John cared for Jane sufficiently that he would have married her had her sister Mary not been available" is evidently needed as part of the justificatory basis for the conditional at issue. Such dispositional relationships can and should also be seen as functioning lawfully in these contexts and thereby lay claim to systemic priority. The general approach sketched in these pages can then meet the needs of the situation without difficulty.

3 Issues of Causality

Theorists sometimes envision a close linkage between causality and counterfactuals. But it is easy to go too far in this direction. In this light, volitional counterfactuals are an object of special interest. We typically explain people's actions in terms of their desires. And in this vein we often propound such counterfactuals as:

If Clyde had not desired a beer, he would not have gone to the refrigerator to satisfy this desire.

The background facts here stand as follows:

(1) Clyde desired a beer.

(2) Clyde went to the refrigerator to satisfy a desire of his.

(3) Clyde's going to the refrigerator had no other purpose than getting a beer.

Observe that $[(2) \,\&\, (3)] \vdash (1)$ so that $[\sim(1) \,\&\, (3)] \vdash \sim(2)$. Accordingly, given (3) we obtain:

If Clyde had not desired a beer, he would not have gone to the refrigerator to satisfy this desire.

So far, so good. We have in hand a factual basis for validating that counterfactual.

Observe, however, that this counterfactual does not actually explain Clyde's going to the refrigerator. For nothing we otherwise have on hand—namely, (1) and (3)—enables us to deduce (2). To do so we will need to add:

(4) Clyde realized that the refrigerator is the best place to get a beer.

(5) Clyde had no preponderant counteracting reason against satisfying his desire for a beer.

Now we have:

(1) & (3) & (4) & (5) ⊢ (2)

Here (3) is the only premise common to our several arguments. A general point is accordingly at issue. Volitional explanations and volitional counterfactuals pivot on purpose-restrictive presuppositions along exactly these lines.

Note moreover that such volitional explanations and counterfactuals proceed entirely within a belief-desire model of human agency. Neither in the premises nor in the *modus operandi* of the reasoning is there any reference to causal operations. Actual causality as such is not essential to counterfactual cogency.

Again, the case of backtracking conditionals serves to indicate that the connection established by counterfactuals is one of assertibility rather than causality. Thus let it be the rule that the game is always played on Sunday afternoon provided that it does not rain. So in the circumstances we can say:

Since it rained last Sunday, the game was not played.

And we can also appropriately say:

If it had not rained last Sunday (as indeed it did) then the game would have been played.

In such circumstances the step forward, as it were, from rain to nonplay is perfectly practicable. However, it is also correct in the circumstances to say:

Since the game was played two Sundays ago, there was no rain then.

And (supposing that it *did* rain then) we could also say:

If the game had been played two Sundays ago, it would not have rained then.

But now of course there is no question of causality (of "backwards causality" as it were). It makes no sense to say that play *causes* nonrain—it does no more than *evidentiate*.

The point here is that the issue of appropriate conditionality is nonprejudicial vis-à-vis causality in that perfectly proper counterfactual conditionals can be in place on grounds of assertibility connections even in the absence of any corresponding causal connection. There is no causal connection among the days of a week; and it is still perfectly in order to say: "If today were not Tuesday, yesterday would not be Monday."

We must in these contexts take care to distinguish between *causal* connections and merely alethic connections of truth status. If I am sad today because (and only because) my friend died yesterday, then it is correct to say "If I were cheerful today (and not sad), then my friend would not have died yesterday." Given that the death of a friend causes sadness for a considerable period this is indeed so. But of course this should not be construed to stake the weird claim that my present sadness caused that past death—that if only I managed to cheer up, my friend would be restored to life. In these contexts of truth connection one must be careful to distinguish between the dependent and the independent variables.

When dealing with counterfactuals in a deterministic world it is appropriate to prioritize laws over occurrences. But what of an indeterministic world? Consider the situation of a stochastic fork in the road, a juncture at which a moving object can (indeterminately) go either up or down (say with 50:50 probability):

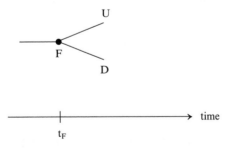

Let us suppose that in the case actually at hand, the object has moved upward. And now consider the counterfactual:

If the outcome had been different (i.e., arrived at D rather than U) then a difference would have been observable at some time prior to t_F.

In the circumstances under supposition this is clearly false. On the other hand, the counterfactual

If the outcome had been different (i.e., at D rather than U) then a difference would have been observable at every time after t_F

would be true. And this also goes for the counterfactual:

Even though the actual outcome of the experiment was U, this outcome was not necessary: D could perfectly well have been the outcome.

As the reader can readily verify, even in such stochastic (and thus indeterministic) situations, the standard procedure of counterfactual analysis based on lawfulness prioritization is unproblematically operable.

4 Modalized Counterfactuals

Modal propositions also fit seamlessly into the fabric of our analysis of the present analysis of counterfactuals. Thus consider:

Even if the match had been struck (which it was not) then it might (still) not have lit, since it could have been wet.

This occurs in a context where we have:

(1) The match was not struck. ($\sim S$)

(2) The match did not light. ($\sim L$)

(3) Wet matches do not light when struck. $(W \& S) \rightarrow \sim L$

(4) Possibly: the match was wet. ($\Diamond W$)

We now reason as follows:

A. S by assumption

B. $\Diamond (W \& S)$ from (A), (4)

C. Possibly: $\sim L$ from (B), (3), since $p \rightarrow q$ yields $\Diamond p \rightarrow \Diamond q$

This suffices to establish the counterfactual. And note that this route to $\sim L$ prescinds from the (2)-given fact "Actually: $\sim L$," thus validating that the use of "still" is the statement of that counterfactual.

Again, consider, for example, a counterfactual of the form:

If p were so (which it isn't), then q might be so.

For concreteness let us instantiate this as:

If Smith were in England, then he might be in London.

Now let it be that this conditional arises in circumstances where we hold the following beliefs:

(1) Smith is not in England.

(2) If one is in England, one might be in London.

And now project the following assumption:

Not-(1): Smith is in England.

With (1) negated as per this assumption, we can use (2) to infer: "Smith might be in London." On this basis, our initial counterfactual follows straightaway. (Of course this reasoning is viable only in belief contexts that fail to place Smith in English locations outside London.)

As this example illustrates, modality-involving counterfactuals can be accommodated by exactly the same process of reasoning that has been employed throughout.

1 *Modus ponens*

Modus ponens argumentation has the format of moving inferentially from the premises p and $p \Rightarrow q$ to the conclusion p. This clearly will not work for counterfactuals: we obviously cannot move on from p and $p \{\mathbf{B}\} \mapsto q$, seeing that $p \{\mathbf{B}\} \mapsto q$ will only obtain when $\sim p$.

On the other hand, since $p \{\mathbf{B}\} \mapsto q$ requires that $(\sim p + (\text{parts of } \mathbf{B})) \vdash q$, we do have:

$\sim p$
$p \{\mathbf{B}\} \mapsto q$
certain accepted beliefs

q

This sort of argumentation is the nearest viable analogue to *modus ponens* that prevails with counterfactuals. But even to call it so—that is, an "analogue"—is something of a stretch.

2 Negation and Denial

When p is a counterfactual hypothesis we will want to have it that (for arbitrary $\mathbf{B}$ with $\sim p \in \mathbf{B}$): $p \{\mathbf{B}\} \mapsto (q \vee \sim q)$. After all, $q \vee \sim q$ (and indeed any theorem of logic) follows from *any* set of premises—even the null set. So far, so good. But of course we will not in general have:

$$(p \{\mathbf{B}\} \mapsto q) \vee (p \{\mathbf{B}\} \mapsto \sim q)$$

The negation of both of these counterfactuals readily obtains.

With counterfactuals as with substantive conditionals we must distinguish between a contradictory *negation*:

$\sim(p \{\mathbf{B}\}\!\mapsto q)$

and a contrary *denial*:

$p \{\mathbf{B}\}\!\mapsto q.$

Recall that our basic construal of counterfactual conditionalization $\{\mathbf{B}\}\!\mapsto$ is such that:

$p \{\mathbf{B}\}\!\mapsto q$ iff $(p + \mathbf{B}^*) \vdash q$ for all p-suitable consistent subsets $\mathbf{B}^*$ of $\mathbf{B}$.

In light of this specification, the negation of a counterfactual and its denial are very different things. Whereas the denial of a counterfactual is itself a counterfactual, the negation of a counterfactual is no counterfactual at all.

To illustrate the process of denial, consider the counterfactuals:

If you had been there (which you weren't), then you
(a)   would have seen the president
(b)   might have seen the president
(c)   would have had to have seen the president,

which answer to the format, respectively, of

$p \{\mathbf{B}\}\!\mapsto q$

$p \{\mathbf{B}\}\!\mapsto \Diamond q$

$p \{\mathbf{B}\}\!\mapsto \Box q$

Their denials are themselves counterfactuals:

$p \{\mathbf{B}\}\!\mapsto \sim q$

$p \{\mathbf{B}\}\!\mapsto \sim\Diamond q$ or equivalently, $p \{\mathbf{B}\}\!\mapsto \Box\sim q$

$p \{\mathbf{B}\}\!\mapsto \sim\Box q$ or equivalently, $p \{\mathbf{B}\}\!\mapsto \Diamond\sim q$

That is, we have the respective denials

(Even) If you had been there, you
would not have seen the president,
could not have seen the president,
might possibly not [or: *need not*] have seen the president.

In each case, this affirms something decidedly stronger than merely negating the original conditional by saying that it fails to hold. For, by contrast, the negations of our three theses are simply contentions that they do not obtain. And this is something else again. Now the three theses of the form:

It is not the case that if you had been there (which you weren't then you) . . .

will come to

$\sim(p \{\mathbf{B}\}\!\mapsto q)$

$\sim\!\Diamond(p \{\mathbf{B}\}\!\mapsto q)$

$\sim\!\Box(p \{\mathbf{B}\}\!\mapsto q)$

And such *denials* of counterfactuality have themselves no element of counterfactuality: it does not take counterfactuals to negate counterfactuals. There is nothing conditional about the negation of a conditional, and this general situation holds for counterfactual conditionals as well.

Not only can counterfactuals be rejected by way of denial and negation, but they can also be rejected as effectively meaningless. Thus observe that if $\mathbf{B} = \{p\}$, the supposition of $\sim p$ would leave us empty-handed, so that $\sim p \{\mathbf{B}\}\!\mapsto q$ will obtain only when $\vdash q$. However, let us add to $\mathbf{B}$ only the p-consequence $p \lor q$, so that now $\mathbf{B} = \{p, p \lor q\}$. Then the supposition of $\sim p$ would leave $p \lor q$ and thus q intact *for arbitrary q*. It is thus fortunate that $p \lor q$ cannot be salient in the presence of p—being comparatively less informative as a logical consequence thereof—so that the initial $\mathbf{B}$-set of purportedly salient beliefs is improperly formed.

The improper character of a certain counterfactual reflects the inappropriateness of the corresponding "what-if" questions. A good example is afforded by the "What-would-you-say-if" question beloved by speculative theorists: if pigs could fly, if flowers could talk, if people's memory banks could be exchanged by electronic devices, and so on. (Section 5 of chapter 14 will return to these issues.)

3 Contraposition

Argumentation by contraposition moves from $p \Rightarrow q$ to $\sim q \Rightarrow \sim p$. It is clear that this fails for counterfactual conditionalization. Consider the counterfactual:

If John were not in Paris, then he might possibly be in Rome.

This clearly does not yield:

If John were not possibly in Rome, then he would be in Paris.

And so we do not have:

$p\,\{\mathbf{B}\}\!\!\mapsto q$ entails $\sim\!q\,\{\mathbf{B}\}\!\!\mapsto \sim\!p$

For the first counterfactual is based on:

$(p + \mathbf{B}^*) \vdash q$ for all p-suitable $\mathbf{B}^*$-sets,

whereas the second requires:

$(\sim\!q + \mathbf{B}^*) \vdash \sim\!p$ for all $\sim\!q$-suitable $\mathbf{B}^*$-sets.

Thanks to the potential difference between the B^*-sets at issue, the transition is just not going to work out in general. Thus consider:

(Even) if Tom had asked for her hand, she would have married Bob (anyway).

There is no viable way of inferring a contrapositive thesis here, for from the antecedent

If she had not married Bob, then . . . ,

there is nothing that we can cogently conclude about Tom's having asked for her hand. Counterfactuals just will not in general contrapose.

4   Disjunction

It is not the case that $p\,\{\mathbf{B}\}\!\!\mapsto (q \vee r)$ entails that either $p\,\{\mathbf{B}\}\!\!\mapsto q$ or $p\,\{\mathbf{B}\}\!\!\mapsto r$—as we have already seen in the special case of $q = \sim\!q$. Thus we will clearly have neither

If π were an integer, then it would be odd

nor

It π were an integer, then it would be even.

But of course

If π were an integer, then it would be either odd or even

is unproblematically correct.

Moreover, with counterfactuals we do not have:

If $p\,\{\mathbf{B}\}\!\!\mapsto q$ and $r\,\{\mathbf{B}\}\!\!\mapsto q$, then $(p \vee r)\,\{\mathbf{B}\}\!\!\mapsto q$.

For the fact that q obtains relative to the p-suitable subsets of **B** and also relative to the r-suitable subsets of **B** does not assure its obtaining relative to the $(p \vee r)$-suitable subsets of **B**. Thus, consider:

If Bizet were Italian, then Bizet and Verdi would be fellow countrymen.
If Verdi were French, then Bizet and Verdi would be fellow countrymen.

∴ If Bizet were Italian or Verdi were French, then Bizet and Verdi would be fellow countrymen.

The conclusion is certainly unacceptable here, seeing that its antecedent would be satisfied by realizing *both* disjuncts, while its conclusion would then be false.

5   Counterfactuals as Nonmonotonic

A cardinal feature of inductive reasonings is that they involve best-fit considerations.[1] This means that inductive reasoning is nonmonotonic in that we can have it that p evidentiates q, while nevertheless $p \& r$ does not. Additional information can destabilize an inductive inference. (Given that someone is a civil engineer it is plausible to infer that the person is male, but given that the person is both a civil engineer and a member of the Daughters of the American Revolution this conclusion is no longer warranted.)

And exactly the same holds for counterfactuals: they too are nomonotonic. For example:

If that fixture were a Tiffany lamp (which it isn't), then it would be worth a lot of money

becomes destabilized when we shift to

If that fixture were a Tiffany lamp (which it isn't) and were broken into a thousand pieces, then it would be worth a lot of money.

Examples of this sort abound. Contrast:

If I had put sugar in the tea then it would have tasted fine

with

If I had put sugar and cayenne pepper in the tea, then it would have tasted fine.[2]

Or again, contrast:

If you greet him, he will answer politely

with

If you greet him with an insult, he will answer politely.

Clearly, the reason why the monotonicity-reflective

If $p \vdash q$ then $(p \& r) \vdash q$

works in deductive contexts is that here there is no enthymematic gap between p and q which requires the addition of further material that may or may not be forthcoming—as per a stipulation of normalcy or of "all things equal" in the case of inductive reasonings. But with counterfactual conditionals we do *not* have:

If $p \{\textbf{B}\} \!\!\rightarrow q$ then $(p \& r) \{\textbf{B}\} \!\!\rightarrow q$.

For $(p + \textbf{B}^*) \vdash q$ for all p-suitable $\textbf{B}^*$ does *not* yield $([p \& r] + \vdash q$ for all $(p \& r)$-suitable $\textbf{B}^*$, in view of the potential difference between the $\textbf{B}^*$-sets at issue. Thanks to best-fit considerations, these two propositional sets need not have any logical relationship to one another but can go their own separate ways.

These considerations make it clear that counterfactual reasoning resembles inductive reasoning in that here too new information can destabilize earlier findings. Nonmonotonicity plays an analogously actual role in both of these inferential domains.

6   Transitivity Failure

Counterfactuals also fail to be transitive. Thus, consider:

If Bizet and Verdi were both Germans, then they would be fellow countrymen.

If Bizet and Verdi were fellow countrymen, then either Bizet would be Italian or Verdi would be French.

Both of these counterfactuals qualify as acceptable and appropriate. But

If Bizet and Verdi were both Germans, then either Bizet would be Italian or Verdi would be French

is clearly false.

Again, the issue is one of general principles. For consider:

$(p + \mathbf{B}^*) \vdash q$ is predicated on the suitability of $\mathbf{B}^*$-sets in respect of p.

$(q + \mathbf{B}^*) \vdash r$ is predicated on the suitability of $\mathbf{B}^*$-sets in respect of q.

But

$(p + \mathbf{B}^*) \vdash r$ is predicated in the suitability of the $\mathbf{B}^*$-sets in respect of p.

And here the circumstance that conceptual suitability of the $\mathbf{B}^*$-sets for different propositions will in general differ means that the linkage needed for transitivity is broken.

7 Further Eccentricities

Many of the logical principles that govern implications can fail when applied to counterfactual conditionals. So at this point, the handwriting is on the wall: counterfactual conditionalization as represented by $\{\mathbf{B}\} \rightarrow$ simply does not qualify as a mode of implication. The failure of *modus ponens*, of transitivity, of monotonicity, and so on make all too clear that it is not an implication relationship that is at issue with this particular mode of conditionalization.

8 Excluded Middle

Counterfactual theorists have debated the status in this domain of what has come to be called the Law of Conditional Excluded Middle (LCEM):[3]

EITHER If p, then q OR If p, then not-q

This "law" clearly holds for material implication ($\supset$) and fails for various other modes of implication such as deducibility ($\vdash$) and strict implication ($\prec$). Moreover (as was noted in section 2 above) it does not hold for counterfactual conditionalization ($\{\mathbf{B}\} \rightarrow$) as we understand it here.

Robert Stalnaker has endorsed and David Lewis rejected the tenability of this principle for counterfactuals—each with plausible-looking argumentation.[4] But in each case, their approach to counterfactuality has been different from the one I offer here, and it is clear that on the present construction of counterfactuals this law will not, in general, hold. This is so for two reasons, among others.

First, p can be entirely irrelevant to the issue of q versus not-q, so that *both* counterfactuals fail to obtain. Thus, for example, we will have neither

If Caesar had not crossed the Rubicon, then Napoleon would not have won at Waterloo

nor

If Caesar had not crossed the Rubicon, then Napoleon would have won at Waterloo.

Both conditionals alike are simply untenable on the present account because the requisite inferential relationships fail.

Second, it could be the case that the body of available information **B**—albeit relevant to q—is simply insufficient for deciding the question of q versus not-q, and adding p does not suffice to settle the matter one way or the other. Once the epistemic factor of an enthymematic recourse to beliefs is introduced into the construal of counterfactual conditionals, the indeterminacy characteristic of failures of LCEM becomes unavoidable.

On the other hand, LCEM's cousin

If p, then EITHER q OR not-q

does hold for all modes of conditionalization, the counterfactual included.

9   Complex Cases: Quantifiers

Just as is the case with ordinary conditionals, counterfactuals can connect not only completed propositions but propositional functions as well, so that the phenomenon of "quantifying in" comes into play. An instructive example of this phenomenon is afforded by the counterfactual contention: "Somebody would have made Queen Elizabeth I a good husband," which we shall here construe as:

There is someone who, had Queen Elizabeth I married him, would have been a good husband for her.

Symbolically, this comes to:

$(\exists x)(M(E, x) \{\mathbf{B}\}\mapsto H(x, E))$

Or again, consider such examples as:

If something that isn't a dog were a dog, then it would be a canine, or symbolically:

$(\forall x)(\sim Dx \rightarrow (Dx \{\mathbf{B}\} \mapsto Cx))$

Even though in fact some *F*s are *G*s, nevertheless if no *F*s were *G*s, then anything that is an *F* would not be a *G*.

$(\exists x)(Fx \& Gx) \& (\sim(\exists x)(Fx \& Gx) \{\mathbf{B}\} \mapsto (Fy)$

$([Fy] \mapsto \sim Gy)$

Such quantified conditionals cannot, of course, be recast in the $p \{B\} \mapsto q$ format since they concern interrelated propositional functions rather than separable self-sufficient propositions.

It is easy to see, however, that such conditionals can be unproblematically accommodated through the treatment of counterfactuals proposed in these chapters. Take, for example, the first of the just-mentioned pair of conditionals. Here we have as beliefs:

(1) All dogs are canines: $(\forall x)(Dx \supset Cx)$.

(2) There are non-dogs: $(\exists x)\sim Dx$.

In virtue of (2) we can revert to the idea of quantificationally geared schematic pseudo-constant, familiar from chapter 3 above. Accordingly we let it be that there is something—*anything*—that is not a dog, so that we have a *u* for which $\sim Du$. And we then suppose that (contrary to fact) this *u* were a dog, so that we suppose:

Du

Then by (1) we obtain *Cu*. This yields:

$Du \{\mathbf{B}\} \mapsto Cu$

And now by generalizing on that universally unrestricted *u* we obtain:

$(\forall x)(Dx \{\mathbf{B}\} \mapsto Cx)$ QED.

Again, consider the counterfactual:

If a man were perfect (which no man is), then he would not be human.

$(\forall x)((\text{man}(x) \& \text{perf}(x)) \{\mathbf{B}\} \mapsto \sim\text{hum}(x))$.

We here have the following beliefs:

(1) No man is perfect.

(2) All men are human.

(3)   No human can be perfect.

And now we suppose:

Not-(1), that is, some man is perfect.

We must, of course, drop (1) in the face of our assumption of not-(1). But
the trio not-(1), (2), (3) is still inconsistent, so either (2) or (3) must go.
Now we may assume that "man" in this discussion means *member of the
biological species Homo sapiens*, and "human" means *finite intelligent
earth-dweller*. So whereas (2) is an empirical generalization, (3) is effectively a conceptual truth. Accordingly, (3) takes priority over (2), and
that supposedly perfect man will not be human by virtue of (5) and (3).
Then, generalizing on our generic man, we than obtain the individual
counterfactual. As this illustration indicates, situations of quantifying-in
are readily associated within the present analysis of counterfactuals.

10   Complex Cases: Probabilities

Probabilistic counterfactuals pose no real obstacle to the presently contemplated approach. Thus suppose an urn with three balls, two red and
one green. Then a ball drawn at random would probably be red. The
question is whether we can now maintain:

If one of those red balls were green, then a ball drawn randomly would
probably be green.

Let the salient beliefs in this situation be:

(1)   Ball 1 is red.

(2)   Ball 2 is red.

(3)   Ball 3 is green.

(4)   The urn contains only balls 1, 2, 3.

(5)   Balls 1 and 2 are the same color.

(6)   Balls 1 and 3 are different colors.

We are now instructed to suppose:

Ball 1 is green.

Then we must clearly give up on (1). But we now face a choice between
(2) and (5), or a choice between (3) and (6).

Table 11.1
A Spectrum of Possibilities

Alternative	Ball Colors		
	1	2	3
Retain (2) and (3)	G	R	G
Retain (2) and (5)	G	R	R
Retain (5) and (3)	G	G	G
Retain (5) and (6)	G	G	R

Table 11.1 surveys the resulting possibilities. It is clear that in most cases, most balls would be green, so that the conditional seems appropriate. And of course when we prioritize the relational (5) and (6) over the more narrowly descriptive (2) and (3) we arrive at just this result.

11 Counterfactual Nesting: Counterfactuals with Counterfactual Components

Some writers regard conditionals with conditional components as per $(p \Rightarrow q) \Rightarrow r$ as inherently problematic.[5] However, once one makes up one's mind as to the exact construction of the implication relationship at issue, there should be little difficulty here. For example, if $\Rightarrow$ is material implication ($\supset$), then $(p \Rightarrow q) \Rightarrow r$ simply comes to $(p \,\&\, {\sim}q) \vee r$. On the other hand, if the implication at issue is deducibility ($\vdash$), then the matter is quite different, and $(p \Rightarrow q) \Rightarrow r$ becomes untenable.[6]

Now, in the special case of counterfactuals, it is readily seen that there need be nothing extraordinary or unnatural about the idea of conditionals within conditionals, as we've seen earlier. Consider, for instance, such a counterfactual as:

If it is indeed true that he blushes if embarrassed, then he should be blushing fiercely now.

If the Kaiser had asked Teddy Roosevelt whether he had ever met Jefferson Davis, then if Roosevelt had answered truthfully it would have been in the negative.

There is surely nothing all that bizarre—let alone unintelligible—about these conditionals.

Again, consider:

If the Lord himself had not been on our side (p), then when enemies rose up against us (q), they would have swallowed us alive (r). (Psalm 124)

This conditional has the form:

$$\sim p \ \{\mathbf{B}\} \!\!\rightarrow (q \rightarrow r)$$

Indeed, there can even be second-order counterfactuals along the lines of "If C_1 were a true/appropriate counterfactual (which it isn't), then some other counterfactual C_2 would obtain." For example: "If it were the case that the counterfactual *If I were in London, then I would be in France* obtained—which, of course, it does not—then it would be the case that *If I were in Knightsbridge, then I would be in France.*"

Such second-order counterfactuals can be handled by exactly the same machinery that has been used right along. Specifically the overall situation is as follows. We have as accepted beliefs:

(1)   It is not the case that: If I were in London, then I would be in France.

(2)   I am not in Knightsbridge.

(3)   Knightsbridge is in London.

(4)   Emplacement-in location is transitive: if A is in B and B is in C, then A is in C.

And we are asked to suppose:

Not-(1): If I were in London, then I would be in France.

Not-(2): I am in Knightsbridge.

Note that these suppositions annihilate (1) and (2) but leave (3) and (4) in place. Moreover, when combined with (3) and (4), our suppositions entail "I am in France." These considerations conjoin to validate that complex counterfactual at issue.

Then too, there is nothing all that far out of the ordinary about a counterfactual-containing counterfactual such as:

If he had asked me to help him (which he didn't) then if he had offered to pay me (which he didn't), I would have helped him.

This admits of representation in the form

$$p \ \{\mathbf{B}\} \!\!\rightarrow (q \ \{\mathbf{B}\} \!\!\rightarrow r)$$

Again, such a counterfactual admits of validation by exactly the same process that has been employed throughout.

12 Variant Analyses of Counterfactuals

1 Alternative Approaches to Counterfactual Analysis: Ramsey's Change-Minimization Test

There are several influential approaches to the analysis of counterfactuals that are very different from the belief-derivationist strategy of the preceding chapters. The earliest of these was first proposed in the 1920s by the English philosopher-logician Frank P. Ramsey. It can be effectively encapsulated in the thesis:

A conditional "If p, then q" is acceptable in the context of a body of belief iff accepting q is required by the result of making the minimal changes in the body of beliefs required to accommodate p.[1]

This minimal belief-revision standard for counterfactual conditionals is perhaps less of a specific tactic for handling them than a general strategy, which has been aptly called "the Ramsey test" by William Harper.[2] It is in some ways a forerunner of our present approach since it too proceeds on a doxastic and thus epistemic basis. Yet it is also substantially different—as will soon become evident.

Ramsey's strategy has been formulated by Robert Stalnaker[3] as follows:

This is how to evaluate a [counterfactual] conditional. First, add the antecedent hypothetically to your stock of beliefs; second, make whatever adjustments are [minimally] required to maintain consistency (without modifying the hypothetical belief in the antecedent); finally consider whether or not the consequent is true.

The governing idea comes down to the thesis

When $\sim p \in B$, then $p \{B\} \mapsto q$ iff $q \in B^-$, where this belief-set B^- is the *minimal revision* of B required to accommodate p self-consistently.

Here B is, as usual, the entire set of belief commitments at issue.

However, the unfortunate fact of it is that the simple-sounding process of belief revision comes to shipwreck on Burley's Principle. Thus suppose I believe p. I then have little choice but to include among my beliefs both p-or-q and p-or-not-q, where q is any arbitrary proposition whatsoever. Let us now consider a conditional of the form

If not-p, then ...

There are then two different alternative minimal readjustments (viz., dropping $p \vee q$ or dropping $p \vee {\sim}q$), and they yield squarely conflicting conditionals:

If ${\sim}p$, then q.

If ${\sim}p$, then not-q

So when supposing ${\sim}p$, the minimal revision process appears to commit us to both q and not-q (for *arbitrary* q).

Again, consider the situation that arises when our beliefs relate to the position of x in a tic-tac-toe framework arrayed as follows:

That is, we believe that x is in the column-row position $(1, 2)$, in committing ourselves to a multitude of beliefs, specifically including:

(1) x is at column-row position $(1, 2)$.

(2) x is not in the first row.

(3) x is not in the third row.

(4) x is not in the second column.

(5) x is not in the third column.

(6) x is not on a diagonal.

Now let us assume x to be elsewhere. Then even after we remove (1) from the list, the rest of those beliefs suffice to yield it back. As ever, Burley's Principle leads us into paradox here. At the very least, two of those other beliefs will have to be sacrificed. But this can, of course, be done in various ways, and all of the available alternatives seem equally qualified. The problem is that there simply is no such thing as a uniquely acceptable minimal revision. Certainly, Ramsey himself did not offer much guidance as to how that minimally revised belief-set B^- is to be formed and left working out the idea of a minimalistic revision as an exercise for the reader. But this is a problem that faces substantial obstacles:

1. Minimality requires a size comparison, but what is to make one revision greater or lesser than another given the unending potential for the internal complexity of what is involved?

2. Minimality becomes impracticable in slippery-slope situations where additional steps towards enhanced differentiation are always possible.

3. Can the idea of a "minimal assumption-accommodating revision" of a belief set be implemented at all? Is minimality something we can actually realize here?

As regards point (1), let it be there are three beagles in the yard and I stipulate "Assume there are two beagles." How are our change-minimizing deliberations to proceed here? Are we to annihilate one beagle? (And which one?) Or should we retain "There are three dogs in the yard" and replace that missing beagle with another dog? (And if so, of what species?) Or should one contract the yard a bit and thus eliminate one beagle from it? The mind boggles.

As regards point (2) of the list, consider something like "Suppose the storm had lasted longer than it had." How are we to think of an alternative that differs minimally: in days, in hours, in minutes, in seconds, in milliseconds? Minimizing the change is clearly impracticable.

Finally, as regards point (3), let it be that Bob weighs 182 pounds. So does Tom. They balance on the teeter-totter. But we will now have the (true) counterfactual:

If Bob's weight differed from Tom's, then one of them would rise to the top on the teeter-totter.

With respect to this difference in weight, there simply is no belief system that both realizes the antecedent and differs minimally from the actual situation. In other words, there simply is no *minimal* revision here in the overall family of relevant beliefs.

Yet another objection to Ramsey's minimal belief-revision standard is that it leads to various counterintuitive results.[4] Thus consider:

If that stuffed owl were still alive, then it would not remain on that shelf tonight but would fly off.

Plausible though this conditional is, it cannot be validated by a minimal belief-revision approach. For in view of its minimality, such a revision would surely keep that revivified owl planted firmly on the shelf—asleep, tethered, walking back and forth in a dazed manner, or some such, all of which yield a world that keeps descriptively closer to the actual than the situation contemplated in the counterfactual's conclusion.

Again, the following example is also instructive in this regard. Suppose our beliefs are:

(1)   Rover is a lapdog.

(2)   Rover is a purebred dog.

(3)   Rover is a prizewinning showdog.

(4)   All dogs are canines.

(5)   Rover is a canine.

And now let it be that we are asked to undertake the following supposition:

Not-(5), i.e., suppose that Rover were not a canine.

We must now abandon not only (5) as per instruction, but also either (4), or all of (1), (2), and (3). On a Ramseyesque minimal revision we would presumably be led to abandon (4) and retain all the rest (save (5)) so as to arrive at:

If Rover were not a canine, then not all dogs would be canines.

This does not seem all that plausible. (Observe, for contrast, that the present information-conservation approach with its prioritization of laws requires the retention of (4), thus leading to the counterfactual:

If Rover were not a canine, then he would not be a purebred, prizewinning lapdog.

which is clearly appropriate in the circumstances.)

It is clear that the present approach differs dramatically from Ramsey's. For in contrast to Ramsey's minimal-revision set B^-, our approach does not involve the wider manifold of all relevant beliefs, but only a small handful of immediately salient items. It thus merely calls, for assessing the tenability of counterfactuals, for us to delete some (generally few) members from a manageably small family of relevant beliefs. Moreover, it does not require a foray into the intractable project of minimizing change with respect to the overall manifold of one's beliefs, but requires merely the application of a few rather straightforward prioritization rules. Ramsey's approach relies, in effect, on a process of "imagining" that projects a transformation of one's entire system of belief.[5] The present approach, by contrast, involves no comparably arcane complications, but proceeds by applying a modest handful of generally straightforward rules. For on its basis we need never grapple with the entire manifold of our beliefs regarding the world at large; only a small handful of immediately beliefs need be brought into it.

2 Alternative Approaches to Counterfactual Analysis: Lewis's World Proximity Criterion

Another influential approach to counterfactual analysis that has been proposed in recent years is based on a possible-worlds approach inaugurated by Robert Stalnaker. Addressing the question "How do we decide whether or not to believe a conditional statement?" Stalnaker maintained that although Ramsey was on the right track to begin with beliefs, nevertheless:

The problem is to make the transition from belief conditionals to truth conditionals.... [And here] the concept of a *possible world* is just what we need to make the transition, since a possible world is the ontological analogue of a stock hypothetical belief. [And so] the following is a final approximation to the account I shall propose: Consider a possible world in which A is true and which otherwise differs minimally from that actual world. "If A then B" is true (false) just as case B is true (false) in that possible world.[6]

Stalnaker's proposal was subsequently extended and developed via what might be called the *world-proximity criterion* of David Lewis. In a series of publications in the 1970s, Lewis proposed a semantical, possible-worlds approach to counterfactuals in which their tenability is determined in terms of "proximity" relations among possible worlds.[7] And, at least in Lewis's earlier papers, proximity is to be determined on the basis of descriptive similarity (though in later papers lawfulness comes to play a more prominent role).

Let us suppose that: "They tell me you will depart tomorrow" is true. And now consider the question: Suppose that that italicized word were not *depart*; what would it then be? Clearly in addressing this question we would follow the Lewis principle and seek an alternative that is as close to actuality as possible—thus opting for an alternative along the lines of *leave* or *go*. But this is a rather special sort of situation. It would make little sense to address the question "What if France had not decided to give the Statue of Liberty to the United States?" with the response "Then France would have given some other vast bronze statue by Bartholdi instead," even though this would keep the resulting world maximality similar to the actual one. Only in very special sorts of cases is the Stalnaker–Lewis similarity approach to counterfactuality in order.

There is a substantial difference between Stalnaker's and Lewis's treatment of counterfactuals and that of the presently articulated approach. Theirs is a metaphysical theory based on the invocation of possible worlds, whereas the present theory is epistemic and thus not lumbered with the metaphysical baggage of merely possible worlds and merely possible objects. Such possibilia, whatever their ontology and their problems, are something very different from sets of beliefs, which can straightforwardly, after all, be finite in scope, incomplete, and indeed sometimes even inconsistent.[8]

Moreover, it is clearly damaging to the Stalnaker–Lewis descriptive similarity approach that we would presumably want to say "If John R. Kennedy had not been assassinated on November 22, 1963, then he (and not Lyndon Johnson) would have delivered the State of the Union message in January of 1964." And yet, clearly, a world in which J.F.K. was assassinated on the day preceding or following the Dallas shooting—and thus unable to give that address—looks to be one descriptively far closer and thus more similar to the actual world.

Then too, a world in which there had been one fewer casuality in the Battle of the Bulge is in its way more similar to the actualities of our world than one in which there had been yet more. But the conditional "If there had been fewer causalities in the Battle of the Bulge than there actually were, there would have been just one fewer of them" does not look all that plausible.

The world-similarity approach also encounters other, more substantial problems. To begin with, there are all of the *practical* difficulties of applying this idea. For example, there is the question of where alternative possible worlds are to come from and how we are to get there from here.[9] But even waiving this problem, how is the idea of world-similarity (or "proximity") to be implemented? After all, given that "similarity" splits apart into an endless multiplicity of respects—just as people can be similar in appearance and dissimilar in personality, or the reverse. Given this proliferation of respects, how can the idea of absolute, across-the-board similarity possibly be implemented? And even if respects are taken into account, how could this be done by some stably context-transcendent weighting? Moreover, which mode of similarity is to prevail: the phenomenologically descriptive or the structural? How is one to cope with *different* modes of resemblance?

One may be tempted to respond that it is qualitative description rather than quantitative structure that matters for world similarity. But before taking this line consider the case of three maps of Europe: one actual and two altered alternatives. In one alternative only the colors of the countries are changed. In the other the colors are kept the same but the shapes (i.e., national boundaries) are modified. Which of these changelings is closer to the original map? Here it would clearly be structural rather than phenomenologically descriptive similarity that matters. As such examples indicate, it would be inappropriate to operate with the idea of similarity in any sort of standardized way.

The prime difficulty is that whenever there is a plurality of different respects there is no single coherent way of moving from a multiplicity of feature correlative modes of similarity–proximity–changelessness to a single, comprehensive, overall similarity–proximity–changelessness. Is a possible world where horses have horns closer to ours than is one where rabbits have pouches? Is a possible world where grass is yellow closer to our world than it is to a world where figs are white? Is similarity among people in point of dieting preferences more or less important for world distance than similarity in point of pastimes? A multitude of questions of this sort immediately arises from the comparison of possible worlds, and there is no practicable way of resolving them.

Consider, for example, the counterfactual:

If Theodore Roosevelt had appointed Woodrow Wilson ambassador of Germany, then Wilson would have mastered conversational German.

How can we possibly tell which of the worlds in which Theodore Roosevelt appoints Woodrow Wilson ambassador to Germany are "sufficiently close" to ours? And where on Earth—or elsewhere—are such worlds to come from?

On the other hand, it is hard to see how Lewis's procedure can avoid endorsing the absurd counterfactual:

Even if Shakespeare had never been born, *Hamlet* would nevertheless still be performed on the stage today.

After all, a world with *Hamlet* performances (even if based on a mysteriously found script by some unidentifiable poet) would seem to be more similar to our world than one where this drama was among the missing. The long and short of it is that there is only a proliferation of similarities of respect and no such thing as a synoptically comprehensive, "everything-considered" similarity.[10]

Just as in personal preference theory there is—by Arrow's paradox—no viable way of extracting a meaningful single overall-preferability index from a variety of preferability respects, so in possible-world theory there is no way of extracting a single overall-similarity index from a multiplicity of respect-geared similarities.

Moreover, on a Lewis-style approach, we are asked to identify those possible worlds where the antecedent obtains and which are maximally (or sufficiently) similar to the actual world. This done, we are to check that the consequent (always) obtains there. But it is clear that on this approach it is now extremely difficult to know what to make of such seemingly acceptable counterfactuals as:

If the big bang had not occurred, then physical matter would not exist.

If the physical constants of nature were different, there would be no stable types of material substance.

If cosmic evolution had taken a different course, stars would not have evolved.

In such cases we would surely find it somewhere between difficult and impossible to say what the appropriate antecedent-satisfying possible worlds would be like.

And further sorts of problems arise, even where a single quantitative feature of the world's composition is concerned, in a way that parallels the difficulties afflicting Ramsey's approach. For the Lewis closest-to-reality standard leads to the anomalous result that for every positive ϵ, we would have the counterfactual "If I were over 7 feet tall then I would (still) be under $7 + \epsilon$ feet tall."[11] Such quantitative parameters are bound to create difficulty for a Lewis-style approach.

Analogous difficulties for this approach arise in the context of temporal processes. Thus consider the following situation: At time t I deliberate about whether to do A at $t + t'$ and decide against it. To all plausible appearances we then have the counterfactual:

If I had not decided at t against doing A at $t + t'$, then I would have done A at that time.

But now for any world in which that antecedent is true (viz., in which I decide at t to do A at $t + t'$) and the consequent also (viz., I do A) there will be a world *that is more similar to the real-world situation* in which the consequence is false (viz., I do not do A), namely, one in which I change my mind soon after t though before $t + t'$. On this basis the true-in-similar-worlds approach to counterfactuals is in principle precluded from validating counterfactuals of the indicated sort.[12]

For Stalnaker and Lewis, the analysis of counterfactuals unavoidably requires a foray into the realm of unactualized possible worlds, scrutinizing those—duly similar to ours—in which the antecedent obtains to see whether whatever is not the consequence obtains there as well. And in view of Burley's Principle it is not easy to exaggerate the extent of the difficulty and perplexity that this sort of project involves.

To be sure, as he was eventually confronted with a variety of concrete examples, Lewis, to his credit, recognized such problems and sought to remove them by introducing curative complications. In particular, he grappled with the (ultimately intractable) difficulties of distilling a single overall similarity comparison out of a plurality of similarities in diverse respects. His solution was to have recourse to the idea of combining similarities via an assignment of different weights to different factors, proposing a "system of weights" to implement the similarity relation for the world-closeness semantics for the truth conditions of subjunctive conditionals,[13] the weighting at issue being based on:

1. minimizing large-scale violations of this world's natural laws;
2. maximizing agreement with this world's large-scale facts;
3. minimizing small-scale violations of this world's natural laws; and
4. maximizing agreement with this world's small-scale facts.

While such a move away from phenomenological toward nomic similarity is unquestionably a step in the right direction, it does not suffice to meet the demands of the situation. For the multiplicity at issue with similarity of respect is simply going to recur at higher levels. There just is no viable way of implementing the crucial idea of minimization/maximization here.

3 Some Points of Difference

A closer comparison of the present approach to counterfactual analysis with that of Stanlaker-Lewis is instructive.

Consider drawing a ball from an urn containing 1,000 white balls and one black one, and let it be that a white ball is in fact drawn, as is all too probable. And now consider the counterfactual

If the drawing had occurred with that number of white and black balls interchanged, a white ball would still have been drawn.

There seems to be no way for a Lewis-style world-similarity approach to avert this result.

However, our present approach averts this outcome. Consider the following salient beliefs:

(1)   There are 1,001 balls in the urn.

(2)   1,000 of these are white, not black.

(3)   One of them is black, not white.

(4)   A white ball was drawn.

(5)   The maximally probable outcome resulted.

We are instructed to replace (2) and (3) respectively with:

(2′)   1,000 of them are black, not white.

(3′)   One of them is white, not black.

Given that (1) is a fixity, this forces a choice between (4) and (5). And with generality preference ruling in favor of (5), we arrive at:

If the drawing had occurred with the number of white and black balls interchanged, a black ball would have been drawn.

This is surely the more sensible outcome.

Again, consider the following tic-tac-toe configuration, with an *x* placed as indicated:

And let us now inquire into the prospects of a conditional of the format:

If *x* were not in the first column, then . . .

Clearly Lewis's approach, in looking for the descriptively nearest possible hypothesis-consonant rearrangement, would be led to endorse the counterfactual:

If x were not in the first column, then it would be (somewhere) in the second column.

But as we saw in section 2 of chapter 8, our present approach, with its prioritization of regularities, leads to a markedly different result, namely:

If x, which is actually at $(1, 2)$, were not in the first column, then it would be at position $(3, 2)$.

There could hardly be a more graphic illustration of the methodological difference between the two approaches.[14]

To make the difference of the various approaches more perspicuous, let us suppose the following situation to prevail with respect to what we know regarding the context of two virtually empty boxes into which Smith has put some coins:

(1)   Box 1 is empty.

(2)   Box 2 has two coins in it.

(3)   There are no pennies in Box 2.

(4)   Smith did not put a penny into either of the boxes.

But now suppose not-(4), that is, assume that Smith had also (additionally) put a penny in one of the boxes.

There are, of course, two possibilities at this point. Smith could have put the penny into Box 1 or into Box 2. Thus two counterfactuals come into our range of contemplation:

(i)   If Smith had also and additionally put a penny into one of the boxes, this penny would be in Box 1 because Box 2 has no pennies in it. (Here we drop not only (4) but also (1).)

(ii)   If Smith had also and additionally put a penny into one of the boxes, this penny would be in Box 2, since Box 1 is empty. (Here we drop not only (4) but (2) and (3) as well.)

On Ramsey's account (least change among prevailing beliefs), and also on Lewis's account (determination by the assumption-consistent possible world closest to the actual), the former of these alternatives would presumably prevail, seeing that by leaving (2) through (4) intact it involves a lesser modification of the status quo defined by the indicated specifications.[15]

Now it might seem that this is also the case with the present account, thanks to crediting (3) with generality precedence. But this appearance is misleading because (3) is no sort of a law of nature but simply a matter of happenstance, which thus has the same footing as (1) and (2). And so we lack adequate priority guidance here. Since in both cases the *overall* plausibility status of what is dropped is the same, we can achieve no more than the disjunctive counterfactual:

(iii)   If Smith had put a penny into one of the boxes, then either Box 1 would contain (only) a penny, or Box 2 would have coins in it, only one of which would be a penny.

This, on our present account, is the most and the best that can be realized.

The difference between the present approach to counterfactuals and that of Ramsey and Lewis can also be illustrated by considering the situation in which we are confronted by the series:

01010101010101 . . .

To all visible appearances this series answers to the rule "Alternate 0s and 1s." But now consider the what-if question:

If that first member of the series were 1, what would the second member be?

With either Ramsey's belief conservation or a Stalnaker-Lewis-style world-similarity approach, we are now embarked on a problematic course. By stipulation, the first member now has to be a 1. But, of course, there remains the prospect of keeping all the other series members as is, and clearly no other alternative arrangement can possibly be closer to the real-would situation. And so on both approaches alike we arrive at:

If that first number were a 1, the second number would (still) be a 1.

Given their insistence on descriptive (phenomenological) real-world similarity, both theorists would have to endorse this conditional.

The present approach, however, does not take this line. In prioritizing general rules over descriptive details we would arrive at:

If that first member were a 1, the second member would be a 0.

And this is surely the more plausible upshot.

The crux here is that whereas those alternative approaches are geared to *descriptively phenomenological approximation* to reality, the present nomologically oriented approach strives for an approximation to reality through *informativeness conservation*.[16]

To be sure, Lewis (in his 1979 paper) eventually recognized the need for law prioritization. However, this departure from his earlier phenomenological approach still rides roughshod over the difference between conceptually grounded laws of logic and language on the one hand, and laws of nature on the other. And it does not reckon with the fact that natural laws themselves can—though applicable always and everywhere—nevertheless be of very different levels of fundamentality. Be this as it may, the gravest objection to the Stalnaker-Lewis possible-worlds approach is that there is no well-defined and operationally effective way of coming to grips with nonexistent possible worlds. Once we introduce hypothetical modification into the fabric of the real we have no practicable way to effect the totality of necessary readjustments.[17]

Considerations along the indicated lines show that the three approaches to counterfactual analysis considered here are in fact predicated on asking quite different questions:

I. *Ramsey*   What is it that obtains in the overall belief system that, while introducing the hypothesis, also includes as much as possible from among our actual beliefs? (Belief-retention standard.)

II. *Stalnaker-Lewis*   What is it that obtains in the possible world arrangement that, while compatible with the hypothesis, comes descriptively closest to (and thus most closely resembles) the arrangements of the real world? (World-proximity standard with proximity construed sometimes descriptively [early Lewis], sometimes nomologically [later Lewis].

III. *Rescher*   What is it that obtains when one curtails the contextually salient beliefs in a way that both assures compatibility with the hypothesis and optimizes the retention of prehypothesis information in the light of the apposite prioritization-standards? (Information-conservation standard.)

The specific implementation of these approaches shows that these are different questions that, as our examples show, have different answers and lead to different results in different cases—results that differ decidedly in their intuitive acceptability.

To summarize: The current scene affords three prime alternatives for counterfactual analysis: Ramsey's minimal revision approach, Lewis's world-similarity maximization approach, and the present book's optimal resystemization approach based on a systematic reconstruction of our belief system using the principles of *saliency* (issue-significance) and *prioritization* (informativeness-precedence). And the claims to superiority of that third alternative reside in two principal considerations:

1. Unlike the alternatives, it does not make virtually unrealizable demands (surveying possible worlds, recasting entire belief systems) but only requires scrutinizing a handful of immediately relevant beliefs.

2. As many concrete examples show, it more adequately accommodates the presystematic acceptability and inductive appropriateness of counterfactual inferences.

4 Can Probabilities Help?

Yet another variant approach to counterfactuals has been proposed, one that proceeds on the basis of probability.[18] It is predicated on having at our disposal a manifold of quantified conditional probability values of the form $pr(p/q) = v$ for the relevant propositions. The idea now is to assess what would have happened if some falsehood obtained by looking at the issue in terms of conditional likelihoods relative to it.

Consider a simple schematic illustration of the operation of this approach. Suppose just three (otherwise independent) propositions are at issue. Then such a schedule of probabilities can be represented as a distribution of likelihoods across the entire possibility-spectrum as per table 12.1. Now let it be that the actual situation is as per case 3 so that p & $\sim q$ & r represents the truth of the matter. And let us then introduce a counterfactual assumption and inquire into what follows, as per the question:

Table 12.1
A Sample Distribution of Likelihoods

	Case			
	p	*q*	*r*	probability
1.	+	+	+	2/16
2.	+	+	−	1/16
3.	+	−	+	3/16
4.	+	−	−	1/16
5.	−	+	+	4/16
6.	−	+	−	2/16
7.	−	−	+	2/16
8.	−	−	−	1/16

Note: The likelihoods of the last column are purely illustrative.

If not p, then what?

On the postulated basis, the presently contemplated probabilistic procedure would proceed by examining the cases in which the projected hypothesis is realized, thereupon *accepting those conclusions that obtain in
all of the maximally likely (or in all of the sufficiently likely) cases.* Thus
in the present case we could either proceed on a maximum-likelihood
standard to arrive at case 5, and thus have:

If not-p, then $q \& r$.

Alternatively, we could proceed on a sufficient-likelihood standard, thus
eliminating the minimal-probability cases 2, 4, 8 and arriving disjunctively at cases 5, 6, 7, thereby validating:

If not-p, then $q \vee r$.

Such an approach may seem promising from a formalistic point of
view. Its fatal flaw, however, lies in the impracticability of implementing
such a procedure. In particular, it leaves unanswered how we could ever
come to grips on such a probabilistic basis with conditionals like:

If Columbus had not discovered America, somebody else would have.

For one thing, the project of surveying that entire spectrum of all theoretically available hypothesis-conforming possibilities becomes hope-
lessly formidable in any real-life application. But more serious yet is the
question of those probability values. To be sure, the calculus of probability tells us how to conjoin those probabilities once we have them.
But where are they to come from and how are they to be determined?
In the realistic cases that concern us there is no hope securing the needed
information.

   Other serious problems also arise. The whole project bears the ominous
onus of a venture in elucidating something obscure in terms of something
that is even more obscure than the target of our initial concern. We want
to know if it is in order to answer questions of the form

If x were the case (which it isn't), then would y be so?

But in addressing this question we are sent off on a treasure hunt for
answers to various questions of the form:

If x were the case (which it isn't), then how likely would it be for y to be
so?

It does not take much to see that proceeding in this way is to resolve one difficulty in terms of another that is yet more deeply problematic.

But perhaps the most telling argument against a probabilistic treatment of conditionals is that the features that standardly define conditionalization as such (*modus ponens*, transitivitiy, and the rest) generally fail to hold for probabilistic relationships.[19] Accordingly, the approach to conditionals taken here takes an altogether different line in its insistence on coordinating conditionality with inferential processes.

13 Historical Counterfactuals

1 Two Types of Historical Counterfactuals

The primary aim of historical inquiry is to elucidate the past—to describe and explain the course of past events. In *describing*, we are, of course, engaged in a strictly factual discussion. Here there is—or should be—little room for fanciful speculation: Leopold von Ranke's insistence that the historian's concern is with "how it actually was" (*wie es eigentlich gewesen war*) stands paramount. However, in *explaining* a course of historical events, it transpires that counterfactuals are an almost indispensably useful resource.

In this regard, the distinction between falsifying and truthifying causal counterfactuals is particularly significant in the context of historical issues:

Falsifying case If something-or-other—which actually did happen—had not happened, then certain specifiable consequences would have ensued. (Example: "If the ministers of George III had not taxed the Colonies, the American Revolution would have been averted in 1776.")

Truthifying case If something-or-other—which did not actually happen—had happened, then certain specifiable consequences would have ensued. (Example: "If the American Colonies had remained subject to Britain, control of the British Empire would eventually have shifted from London to North America.")[1]

With that first, falsifying example we have a situation in which our salient beliefs are as follows:

p: George III's ministers taxed the American Colonies.

q: The American Colonies revolted in 1776.

Without p, we would have not-q: not-p ⇒ not-q: The American Revolution of 1776 was in substantial measure a response to George III's ministers taxing the colonies.

We are now asked to suppose not-p. Overall, we confront the following situation: Assume not-p in the face of the aforementioned beliefs: p, q, not-p ⇒ not-q. The problem then is that even after dropping p, we must come to terms with the inconsistency of the group: not-p, q, not-p ⇒ not-q. And so here our supposition of not-p constrains a choice between the specifically factual q and the relational not-p ⇒ not-q.

By contrast, in the second, truthifying example, our salient beliefs are as follows:

not-p: The American Colonies did not remain subject to Britain.

q: Control of the British Empire has remained in London rather than North America.

Since q then not-p: q ⇒ not-p: The retention of North America would eventually have disestablished the primacy of London.

And we are asked to suppose counterfactually that p. Thus, here we have the situation: Assume p in the face of those specified beliefs: not-p, q, q ⇒ not-p. The problem then is that even after dropping not-p, we must come to terms with the inconsistency of the group: p, q, q ⇒ not-p. In short, our supposition constrains a choice between the specifically factual q and the relational q ⇒ not-p.

Thus in both cases alike we must choose between a categorical and a relationally conditional fact. And as long as we prioritize comparatively more general relationships (conditional relationships included) over more restrictedly particular claims on grounds of informativeness, we must, again in both cases alike, abandon the former and accept the latter in its place. In their logical structure the two cases are thus very similar, although they differ quite significantly in other respects.

Historical counterfactuals of the falsifying type—"if such and such had not happened"—generally address the *preconditions* for an actual occurrence—its temporally antecedent requisites. They are in general *retrospectively cause-determinative* in nature. And this sort of thing is—or ought to be—grist to history's mill given the explanatory mission of the enterprise. (It should be noted that counterfactuals of regret along the lines of "If only I had not done x but had done y instead, I would not be in the mess I'm in today" are in effect historical counterfactuals of this retro-spectively cause-determinative variety.)

By contrast, historical counterfactuals of the truthifying type—"if such and such had occurred"—generally address the *consequences* of a purely hypothetical event. They are in general *prospective* and *speculatively consequence-determinative* in nature. This sort of thing is historically far more problematic, given history's intimate linkage to the world's actualities.[2]

With both of these types of counterfactuals we enter into the virtual re-ality of a suppositional realm of "what if," but in the second, truthifying case we generally skate on much thinner ice. The unrealism of the former, cause-oriented type of historical counterfactual is less severe than that of the latter, consequence-oriented type. The actual causal course of events now fades into the background, and it is the presumptive consequences of merely suppositional, actually nonoccurent events that now concern us. In sum, while falsifying counterfactuals remain within the orthodox bounds of a reality-based causal analysis—since in looking to the causes of actual eventuations we are still in effect concerned with how things have actually happened—truthifying counterfactuals, by contrast, are by nature usually more speculative and conjectural.

2 Temporal Asymmetry

Several recent writers have deliberated about the time direction at issue with counterfactual conditionals,[3] being concerned that past-directed counterfactuals pose special difficulty. The present approach can come to grips with these issues in a rather straightforward way. Thus con-sider a journey made up of successive daily stages along the following path:

Here there is clearly no difficulty with:

If the traveler were at B today (rather than at C as is actually the case), then he would have been at A yesterday.

By contrast, suppose that the path system were to have a backward-forked pathway structure, so that several routes to a given destination are at issue, as per:

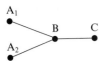

Now we can do no better than:

If the traveler were at B today (rather than at C as is actually the case) then he would either have been at A_1 or at A_2 yesterday.

In other words, the question "Where would the traveler have been yesterday if he were at B today rather than at C?" is now one to which—in the informative circumstances at issue—we simply cannot provide a definitely univocal answer.

On the other hand, suppose that the path system were:

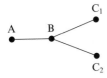

It is now the question "It the traveler were at B today, rather than at A as is actually the case, then where will he be tomorrow?" that we cannot definitely answer.

Since given the role of chance in our world it is structured so that there are fewer causal convergent paths from the past than divergent causal paths to the future, it emerges that past-oriented ("backward-looking") counterfactuals are in a more tractable condition that future-oriented ("future-looking") ones. The situation of counterfactuals in relation to time order is thus destined not to be symmetric. The life of the historian is easier than that of the seer.

3 Examples

Counterfactual historical suppositions occur not only in claims about specific occurrences (events), but also in claims regarding general trends and tendencies (connected courses of events). And, in resolving all such counterfactuals, the rule—as usual—is to prioritize informativeness.

Consider the counterfactual:

If the European component of World War II had been prolonged by six months or more, the United States would have used the atom bomb against Germany.

This counterfactual arises in a context where the following propositions qualify as accepted facts:

(1)   The European component of World War II ended in April 1945, rather than substantially later.

(2)   The United States did not have its atomic bombs ready for use until August of 1945.

(3)   The United States did not use the atom bomb against Germany.

(4)   The United States developed the atom bomb primarily with a view to its use against Germany, and would have used it to end the war in Europe if required.

Note that even if we abandon (1) and replace it with its negation we would still have a contradiction on our hands, since not-(1) in combination with (2) and (4) contradicts (3). In effect we are forced to a choice between (3) and (4) as long as (2) remains in place as a fact that is not in question, because the context of the problem set by the supposition of by (1)-abandonment is the history of that war's end, not the history of technology. And so the standard policy of prioritizing generalities over specific events and event complexes leads to the validation of the counterfactual stated above.

Again, consider the counterfactual:

If France had not aided Britain's American colonies during their revolt, they would not have won their independence.

In this context the following propositions may be taken as accepted:

(1)   France aided the American Colonies (by way of financial and naval support).

(2)   The American Colonies won their independence.

(3)   The financial and military resources of the American Colonies were insufficient to conduct a long, seriously contested war against Britain without substantial foreign financial and military support.

(4)   Among the world powers of the day, only France was willing and able to lend the American Colonies such support.

(5)   Only a long, seriously contested war could possibly gain independence for the American Colonies.

Observe that with (1) deleted and replaced by its negation we still have a contradiction: not-(1) when combined with (3) through (5) yields not-(2). In effect, we must choose between (2) and (3) through (5). However, the statements at issue here differ significantly in their epistemic status. Whereas (1) and (2) concern particular historical facts, (3) through (5) concern systemic historical facts. And so, in line with the general policy of informativeness priority, we come to the result that (2) must be abandoned. The counterfactual at issue becomes established on this basis.

Another plausible example of a falsifying historical counterfactual is:

If Wellington had lost at Waterloo, then Napoleon would not have been forced into (immediate) exile in St. Helena.

This admits of the following justifactory analysis, in which our beliefs are:

(1)   Wellington did not lose at Waterloo.

(2)   Wellington and Napoleon were opposed commanders at Waterloo.

(3)   Napoleon went into (immediate) exile.

(4)   Victorious commanders are not forced into immediate exile.

And we are asked to assume:

Not-(1): Wellington lost at Waterloo.

Even if we drop (1) in the wake of the assumption we still have a contradiction. For this assumption together with (2) entails "Napoleon won at Waterloo," and thus also "Napoleon was a victorious commander." And this together with (4) yields not-(3).

Given that (2) is not at issue here, our assumption of not-(1) in effect forces a choice between (3) and (4). And when we prioritize general relationships over particular facts—as is standard in counterfactual situations—we will retain (4) so that (3) must be sacrificed. And on this basis that initial counterfactual becomes validated.

Another plausible example of a truthifying historical counterfactual runs as follows:

If Julius Caesar had not crossed the Rubicon in revolt against the Roman republic, this republic would have endured far longer.

Here we have as accepted facts:

(1)   Caesar crossed the Rubicon in revolt against the Roman republic.

(2)   Caesar's revolt destabilized the republic and rendered it unstable and untenable.

(3)   Without this modality and conditionality the republic would have endured far longer.

We now assume not-(1). Then, by (2), Caesar's action has not (would not have) rendered the republic unstable and untenable, so that, by (3), it would have endured longer. Here the conditional at issue follows simply from the hypothesis plus the limited register of salient facts.

As such illustrations show, maintaining consistency in the case of historical counterfactuals envisions a priority ranking of the by now familiar sort, geared to the principle of fostering informativeness via the successive prioritization of:

• generalizations, "laws," well-established general rules, well-confirmed causal relationships;

• trends, established relations among events, connections among particular eventuations, coordinative either-or or if-then relationships; and

• specific, particular, concrete eventuations.

As ever, large-scale informativeness is the crux for counterfactual reasoning rather than concrete specificity, as is the case with factual deliberation. And so the priority situation in these *speculative* cases is exactly the reverse of the order of evidential security that obtains in the factual setting of *inductive* reasoning, where theory must give way to facts and more far-reaching theories to those that are more particular in their bearing. Specifically historical counterfactuals are in the same boat as others in that they pivot on considerations of rational economy via a principle of the conservation of information.

And so, the crucial point for present purposes is that historical counterfactuals can—whenever plausible—be accommodated readily and naturally by the general process of counterfactual analysis that has been set out in the preceding chapters.

4   Conclusion

To be sure, many historical counterfactuals lie in the region intermediate between the implausible and the cognitively intractable. Historical counterfactuals, like counterfactuals in general, must rest on an enthymematic basis of background belief, and such beliefs may, of course, be disputable and controversial. Thus consider once more the counterfactual:

If Oswald had not shot J.F.K., then no one would have.

The problem with this conditional is that its tenability requires us to have belief-warranted access to a body of information B of a sort that if conjuntively added to "Oswald did not shoot J.F.K." it would warrant the conclusion "No one shot J.F.K." Now, at the very least, this information would have to assure that there was *not*, waiting somewhere in the wings, someone else who would have shot J.F.K. had Oswald not done so. But how in the name of common sense could we ever ascertain this even if it were so? Such a claim may well be beyond the range of secure determinability and is thus bound to be controversial. Replaying the course of history to see how things would have turned out in such a case is usually problematic on grounds of missing or uncertain information. After all, many—and perhaps even most—historical counterfactuals involve issues that suffer from insufficient or debatable data; the sorts of speculations they present are all too often insubstantiable. This, no doubt, is the principal reason why historians have traditionally been reluctant to engage in this sort of discourse.

Alongside the impracticality of a detailed specification of merely possible worlds, that of a detailed specification of alternative courses of history at large also looms. Possible alternative histories are as little amenable to detailed specification as are possible alternative worlds at large. Accordingly, those accounts of historical counterfactuals that turn on specifying merely possible histories[4] are in the final analysis every bit as unsatisfactory as a possible-worlds treatment of counterfactuals. The problem is basically the same on both sides, namely that we simply cannot effect a sufficiently detailed implementation of this idea. Nevertheless, there is no reason to regret this, for an adequate analysis of historical counterfactuals can be provided within the framework of a belief-readjusting approach along the lines of that which has been contemplated throughout these pages. And here only a handful of our actual beliefs need be brought into it. No detailed reconstruction of world history at large is ever required.

14 *Per impossibile* Counterfactuals and *Reductio ad absurdum* Conditionals

1 *Reductio ad absurdum* Reasoning

Let us now turn to a region close to that of logical theory itself, namely inference by *reductio ad absurdum* argumentation—a special case of demonstrative reasoning. What we deal with here is an argument of the following pattern:

From the situation:

(a to-be-refuted false assumption + firmly established facts) ⊢ contradiction

to conclude the denial of that to-be-refuted assumption.

This mode of *reductio* argumentation is thus also based upon introducing a belief-inconsistent supposition. The object of the exercise is to refute a certain thesis in a setting where various other theses are already firmly in place. The argumentation at issue proceeds as follows: Assume that not-p. Show that, in the presence of those prior, already established and accepted theses (let them be $t_1, t_2, \ldots, t_n$), it transpires that this assumption yields some self-contradictory proposition x as a deductive consequence:

$(\text{not-}p, t_1, t_2, \ldots, t_n) \vdash x$

This being so, we are to conclude that p must be accepted because if not-p were the case, an absurd consequence would result in the setting of our (nonnegotiable) commitments. For since all of those mediating theses are accepted as firmly established facts, while not-p is now no more than a tentatively adopted provisional assumption, one must grant priority to those preestablished givens, and accordingly we have to abandon not-p, thus classing p itself as an established fact.

Thus, *reductio ad absurdum* reasoning is a version of reasoning from a belief-inconsistent supposition. But what is now at issue is a process with a very particular sort of end in view: with *reductio* argumentation, the assumption at issue is to be classed as the most—rather than as elsewhere the least—vulnerable proposition: one whose priority is now so low that it must give way in the face of any effective opposition.

With *reductio* reasoning—in emphatic contradistinction to ordinary hypothetical reasoning—assumptions are not sacred but, on the contrary, are to be seen as the potentially weakest link. The operative principle of procedure is: "Restore consistency while preserving at all cost what is already established, sacrificing any mere assumption to this body of preestablished fact," since consistency demands it. Such a rule to the effect that established propositions prevail over mere assumptions, come what may, makes conflict resolution in these *reductio ad absurdum* cases a straightforward business. As usual, we salvage matters by breaking the chain of inconsistency at its weakest link. But in the present circumstances this is going to be the very assumption that is at issue.

Situations of *reductio* reasoning are thus decidedly different from those of ordinary cases of counterfactual deliberation. And here the important larger lesson emerges that the specific *probative context of consideration* will determine the way in which propositional prioritization occurs in matters of counterfactual reasoning.

2 Some Examples of *Reductio* Reasoning

Some examples will further clarify matters. A classic instance of *reductio* reasoning in Greek mathematics relates to the discovery by Pythagoras—disclosed to the chagrin of his associates by Hippasus of Metapontum in the fifth century BC—of the incommensurability of the diagonal of a square with its sides. The reasoning at issue runs as follows:

Let d be the length of the diagonal of a square and s the length of its sides. Then by the Pythagorean theorem we have it that $d^2 = 2s^2$. Now suppose (by way of a *reductio* assumption) that d and s were commensurable in terms of a common unit n, so that $d = n \times u$ and $s = m \times u$, where m and n are whole numbers (integers) that have no common divisor. (If there were a common divisor, we could simply shift it into u.) Now we know that

$$(n \times u)^2 = 2(m \times u)^2.$$

We then have it that $n^2 = 2m^2$. This means that n must be even, since only even integers have even squares. So $n = 2k$. But now $n^2 = (2k)^2 = 4k^2 = 2m^2$, so that $2k^2 = m^2$. But this means that m must be even (by the same reasoning as before). And this means that m and n, both being even, will have at least one common divisor (namely 2), contrary to the hypothesis that they have none. Accordingly, since that initial commensurability assumption engendered a contradiction, we have no alternative but to reject it. The incommensurability thesis is accordingly established.[1]

This sort of proof of a thesis by *reductio* argumentation that derives a contradiction from its negation is a common form of reasoning in mathematics, and is there characterized as an *indirect proof*.

Again, consider another example. Thompson's Lamp Paradox was suggested by the English philosopher James Thompson,[2] who posed the following question:

A lamp has two settings: on and off. Initially it is on. During the next 1/2 second it is switched off. During the subsequent 1/4 second it is switched on. And so on with a change of switch setting over every interval half as long as the preceding, alternating on and off. Question: What is its setting at exactly one second after the start?

This situation gives rise to the following apory:

(1)   A lamp of the hypothesized kind is possible.

(2)   At any given time, the lamp is on or off, but not both. Moreover,

(3)   Physical processes are continuous. A physical condition that prevails at some time within every ϵ-sized interval prior to t, no matter how small ϵ may be, will prevail at t as well.

(4)   Within every ϵ-sized interval prior to $t = 1$ second the lamp is frequently on.

(5)   Within every ϵ-sized interval prior to $t = 1$ second the lamp is frequently off.

(6)   At $t = 1$, the lamp is on. (By (3) and (4))

(7)   At $t = 1$, the lamp is off. (By (3) and (5))

In the wake of this contradiction our basic assumption of (1) has to be abandoned. The possibility of such a lamp with its peculiar mode of comportment is abolished through a *reductio*.

Such illustrations illuminate the contrast between standard hypothetical inference and *reductio* reasoning. Ordinarily in hypothetical reasoning
the assumptions we make are *issue-definitive* stipulations and consequently allowed by fiat to prevail come what may. In *reductio* reasoning,
however, our assumptions are *merely provisional* and must in the end give
way in cases of conflict with established facts. With counterfactual reasoning, assumptions were treated as sacred. Seen as fixed points around
which everything else had to revolve, their priority was absolute. With
reductio ad absurdum reasoning, by contrast, the matter is reversed. Here
assumptions are viewed as merely tentative hypotheses and accordingly
are frail and vulnerable. Our stance toward them is that of a fair-weather
friend who abandons them in the face of difficulties: they stand at the bottom of the precedence-priority scale.

3   *Reductio* Conditionals

While *reductio ad absurdum* argumentation, understood along the preceding lines, is a characteristically distinctive process of reasoning, the consequential arguments at issue here will, as usual, engender counterfactual
conditionals. For we can, of course, simply adjoin a self-contradictory
conclusion to the assumption(s) in question via the conditionalizing "if-
then." Thus: "If the diagonal and sides of a square were commensurable,
then their measurement would yield a value that is both even and un-
even." The conditional makes manifest that in such a case the antecedent
leads to absurdity.

And by closely analogous reasoning we can also envision *reductio*
conditionals whose conclusions are not self-contradictory but rather inconsistent with certain blatantly obvious facts. Some conditionals that
instantiate this sort of situation are:

If that's so, then I'm a monkey's uncle.

If that is true, then pigs can fly.

If he did that, then I'm the Shah of Persia.

What we have here are consequences that are absurd in the sense of being
obviously false. Despite its departure from what is strictly speaking so
construed, this sort of thing should be reguarded as an attenuated mode
of *reductio ad absurdum*.

4   *Per impossibile* Reasoning

Per impossibile reasoning also proceeds from a patently impossible premise and is accordingly closely related to (albeit distinctly different from)
reductio ad absurdum arguments. Here those literally impossible suppositions are not just dramatically but *necessarily* false, thanks to their logical
conflict with some clearly necessary truths, be that necessity logical or
conceptual in nature or mathematical or physical. Thus, such a literally
impossible supposition may negate a matter of (logico-conceptual) necessity ("There are infinitely many prime numbers"), or a law of nature
("Water freezes at low temperatures").

Suppositions of this sort give rise to such *per impossibile* counterfactuals as:

If (*per impossibile*) water did not freeze, then ice would not exist.

If (*per impossibile*) there were only finitely many prime numbers, then
there would be a largest prime number.[3]

If, *per impossibile*, pigs could fly, then the sky would sometimes be full
of porkers.

If you were transported through space faster than the speed of light, then
you would return from the journey younger than you were at the outset.

Even if (*per impossibile*) there were no primes less than 1,000,000,000,
the number of primes would be infinite.

With these *per impossibile* counterfactuals we envision what is acknowledged as an impossible and thus *necessarily* false antecedent,[4] not in
order to refute it as absurd (as in *reductio ad absurdum* reasoning), but
in order to do the best we can to derive the "natural" consequences that
ensue. For now the operative assumption regains its absolute priority,
notwithstanding its recognized impossibility.

The designation *per impossibile* indicates that it is the conditional itself
that concerns us. Our business here is with the character of that consequential relationship. In this regard the situation is quite different from
reductio argumentation by which we seek to establish the untenability of
the antecedent. The point is simply that from $p \vdash$ (contradiction) one can
infer both $\square \sim p$ and $p \rightarrow (p \,\&\, \sim p)$.

In validating *per impossibile* counterfactuals we accordingly proceed on
the same principle as with "normal" counterfactuals rather than with the

special conventions pertaining in *reductio ad absurdum* reasoning. The reasoning at issue here has the following familiar structure:

(impossible assumption + accepted facts) ⊦ impossible conclusion

Once again, such an inferential situation can be enthymematically summarized in a corresponding conditional:

impossible assumption ⇒ impossible consequence

But one now calls it a day, rather than moving on to the flat-out rejection of the antecedent.

Again, consider such counterfactuals as:

If (*per impossibile*) 9 were divisible by 4 without remainder, then it would be an even number.

If (*per impossibile*) Napoleon were brought back to life today, he would be surprised at the state of international politics in Europe.

Such conditionals do not ask us to *imagine* the impossible (no doubt an unrealizable task!) but merely to *suppose* it. Thus in the first case we have the following items of salient background knowledge:

(1)   Every number divisible by 4 without remainder is even.

(2)   Nine when divided by 4 leaves remainder 1.

In the face of the supposition at issue we will have to abandon one of these, and will proceed, as usual, to prioritize generalities over specificities. So it retains (1) and abandons (2), thus arriving at the if-then relationship of the counterfactual at issue by continuing its hypothesis with (1). In effect, the conditional is simply an emphatic reaffirmation of (1).

The burden of *per impossibile* counterfactuals is generally borne by a single factual contention that functions as the enthymematic basis of the conditional. Thus, consider the conditional:

If (*per impossibile*) 2 were an odd number, then some odd integer would have an odd successor.

This is clearly predicated on "the successor of 2 is 3, which is an odd number." And it could be argued that such *per impossibile* conditionals are at bottom no more than a stylistic variant of their factual basis so that they are not genuinely conditional at all but merely flat-out facts in conditional disguise.

However, sometimes what looks to be a *per impossibile* conditional actually is not. Thus consider:

If I were you, I would accept his offer.

Clearly the antecedent/thesis "I = you" is on the face of it absurd. But even the slightest heed of what is communicatively occurring here shows that what is at issue is not this just-stated impossibility but a counterfactual of the form:

If I were in your place (i.e., if I were circumstanced in the condition in which you now find yourself), then I would consult the doctor.

Only by being perversely literalistic could we see the protasis of this conditional as involving an authentic impossibility. And there is a vast spectrum of counterfactual statements where a comparable situation obtains. Thus consider:

If people could have things wholly their way, then . . .

If you had been born to my parents, then . . .

If only he were smaller than he actually is, then . . .

All such "far out" counterfactuals can be analyzed in a way that brings them within the orbit of our present approach. Consider, for example, the above-mentioned conditional:

If I were you, I would consult the doctor.

The background beliefs here include some such assortment as the following:

(1)   You and I are two different people.

(2)   You have a certain medical problem that I don't.

(3)   If I had a medical problem, I would do the sensible thing.

(4)   The sensible thing to do about a medical problem is to consult one's doctor.

(5)   You are not minded to consult your doctor.

We now introduce the supposition "Suppose I were in your shoes." In the wake of this supposition we must jettison (1) and (2). Since my having that medical problem now follows, it emerges from (3) and (4) that I would consult the doctor—exactly the result which that conditional insists. Here the idea that if I were you then I would do what I am antecedently minded to (viz. (3)) rather than what you are antecedently minded to (viz. (5)) provides the pivotal basis for this line of argument.

In validating *per impossibile* counterfactuals we thus proceed—as usual—by principles of priority that keep the systemically more fundamental beliefs intact. The optimal conservation of information is again the key.

5   Limits of Supposition

Even though we are—or should be—prepared to contemplate impossible suppositions for the sake of discussion, there nevertheless remains a limit to how far we can reasonably go in tolerating *per impossibile* claims. This limit is set by the limits of meaningfulness.

Our concepts generally develop against the background of an understanding of how things work in the world (or, at any rate, are taken by us to work). In consequence, these concepts are such that their very viability is linked indissolubly to the experienced realities of this actual world. They are made for everyday use and cannot survive unaltered in the more stressful atmosphere of unrestrained speculation and merely theoretical musings. Someone seeking to "clarify" such concepts by introducing fact-contradicting hypotheses in the interests of theoretical tidiness will in fact simply distort and destroy them.

The contemporary literature of the philosophy of mind is full of robots whose communicative behavior is remarkably anthropoidal (are they "conscious" or not?) and of personality exchanges between people (which one is "the same person"?). But all such proceedings are intrinsically defective. The assumptions at issue call for the suppositional severing of properties and features that normally go together—and do so in circumstances where the concepts we use are predicated upon a certain background of as-things-generally-go "normality." No supposedly clarificatory hypothesis should arbitrarily cut asunder what the basic facts of this world have joined together—at any rate not where elucidating those concepts whose life-blood is drawn from the source of fact is concerned. If we abrogate or abolish this factual framework by projecting some contrary-to-fact supposition—however well-meaningly intended to clarify the issues—we thereby destroy the undergirding basis that is essential to the applicability and viability of these concepts.[5]

Consider another example, John Stuart Mill's critique of any theory of substance that contemplates a nonsensible *substrate* of sensation. Mill writes:

If there be such a substratum, then suppose it is at this instant miraculously annihilated, and let the sensations continue in the same order. How would the *substratum* be missed? By what signs should we be able to observe that its existence had been terminated? Should we not have as much reason to believe that it still existed as we now have? And if we should not then be warranted in believing it, how can we be so now?[6]

But note that Mill's thought experiment turns on our supposing that *"it [the substratum] is ... annihilated and ... [the] sensations continue [unchanged and] in the same order."* This supposition is, on the face of it, absurd. If the nonsensible substrate of sensation indeed is what it is by hypothesis supposed to be in its very nature—namely, that which accounts for the substance and the ordering of our sensations—then the hypothesis we are being invited to make is simply self-contradictory: it makes no sense to suppose the phenomenon in the absence of that which—by hypothesis —supposedly produces it. (It would be like imagining *sun*light in the absence of the sun.) If—as is indeed the case—our standard view of the world is *de facto* a causal one, so that our sensations are taken to have nonsensuous causes, then the prospect of discussing this nonsensuous causal basis without thereby annihilating its sensuous results is simply absurd.

And so a substantial lesson is at issue here. Our experientially based concepts are—and must be—inherently geared to the world's contingent *modus operandi.* They are made into viably integrated units only by the factual arrangements of the world in which they have evolved. Accordingly they are held together by the glue of a substantive view of the empirical facts. Such fact-based concepts have an inner structure in which theoretically separable factors are conjoined in coordinated juxtaposition. They lack the abstract integrity of purely theoretical coherence that alone could enable them to accommodate the demands of fact-abstractive precision. And so, when the very meaning of a concept presupposes certain facts, its explication and analysis clearly cannot—in the nature of the case—suppose that this basis is simply abrogated. The operation of such concepts cannot be pressed beyond the cohesive force of the factual considerations that synthesize them into meaningful units. Once we project a fact-contradicting supposition that abrogates the basis of such a concept, its meaningful employment is automatically precluded. The limits of meaningfulness accordingly set limits to meaningful supposition.

6   A Lesson

With both *per impossibile* and *reductio ad absurdum* reasoning we set out from an "impossible" or "absurd" assumption that stands in flat contradiction with what we know. But there is a crucial *purposive* difference between the two modes of thought. In the case of *reductio* the aim is to *establish* the absurdity at issue by establishing that and how the contradiction in question follows. In the case of *per impossibile* reasoning, however, the impossibility is conceded but waived: the aim is to bring the dire implications of that "impossible" supposition to light. We simply want to show that a certain consequence follows. And the purposive difference between the two projects accounts for the very different ways in which they address the prioritization of hypotheses.

The main point, however, remains that even the sort of eccentric counterfactuals that are at issue with *reductio* and *per impossibile* conditionals can readily be accounted for within the framework of an epistemic-deductive analysis of counterfactual conditionals along the general lines articulated in the present discussion.

1 Problems of Identifying Possible Worlds

By their nature as such, worlds do and must have a definite character. A world is not just any state of affairs,[1] but will have to embody a "saturated" or "maximal" state of affairs-at-large—a state that affairs-in-toto can assume as a synoptic totality that suffices to resolve as true or false any descriptive claim whatsoever. (Unlike the state of affairs expressed by "A pen is writing this sentence," a world cannot leave unresolved whether that pen is writing with black ink or blue.) If an authentic world is to be at issue (be it existent or not) this item must "make up its mind," so to speak, about what features it does or does not have.[2] Any assertion that purports to be about it must thus be either definitively true or definitively false—however difficult (or even impossible) discovering this one way or the other may prove to be for particular inquirers.

And herein lies the rub. For we frail humans can never manage to *identify* such a totality. Consider a state of affairs indicated by such a claim as "The pen at work here is blue." A real pen cannot *just* be blue: it has to be a definite shade of blue—generic blueness will not do. And so a world-description cannot rest satisfied with "There are two or three people in the room"—that world has to go one way or the other. For this reason, alas, no matter how much we say about something, the reality of concrete particulars will go beyond it. As regards those merely possible worlds, we simply have no way to get there from here by descriptive means. To identify a nonexistent world descriptively—as we must since ostension is unavailing here—we have to specify *everything* that is the case regarding it, and this simply cannot be done.

To be sure, the identification of *our* (actual) world is no problem. The matter is simplicity itself. All we need do is to indicate that what is at issue is *this* world of ours (thumping on the table).[3] The very fact of its being the world in which we are all present together renders such an essentially ostensive identification of this world unproblematic. However, the matter is very different with other "worlds" that do not exist. One clearly cannot identify them by an ostension-involving indication that is, by its very nature, limited to the domain of the actual.

With God, no doubt, possible objects can be identified via possible descriptions contemplated by the divine intellect. For even as God has direct knowledge of actualities by intellectual vision, so he has indirect knowledge (*scientia media*, or middle knowledge) of all the possibilia which are, after all, creatures of his own thought.[4] No doubt such a scholastic account can be maintained at the theological level. But for our human dealings with possibilities it is beyond reach. When it comes to dealing with *possibilia* instead of possibilities, all we can say is: "God only knows!"[5]

Authentic world-descriptions are simply not available to finite beings. Their limitless comprehensiveness makes it impracticable to get a descriptive grip on the identifactory particularity necessary for anything worthy of being characterized a nonexistent world. And so from this angle too we reinforce the thesis that the alternative reality of many hypothetical individuals and worlds is bound to deal in abstracta and thereby unable to present concrete and authentic objects. And here the situation as regards possible worlds is, if anything, even more problematic than that of possible individuals.

Seeing that we can only get at unreal possibilities by way of assumptions and hypotheses, and that such assumptions and hypotheses can never succeed in identifying a concrete world, it follows that we can only ever consider such worlds schematically, as generalized abstractions. Once we depart from the convenient availability of the actual we are inevitably stymied regardless the identification of nonexistent particular worlds. Whatever we can appropriately say about such "worlds" will remain generic, able to characterize them only insofar as they are of a certain general type or kind. Possible-world theorists have the audacity to employ a machinery of clarification that utilizes entities of a sort of which they are unable to provide even a single identifiable example.

2 Problems of Strong Possible-World Realism

Possible-world semantics emerged partly in response to the difficulties in explicating talk about unrealized possibilities in terms of families of sentences. For various difficulties arise in determining sets that would do the needed work, and Nelson Goodman argued that it appeared to be in principle impossible. But insofar as the presently contemplated approach is serviceable, such pessimism is not justified; considerations of epistemic priority and precedence provide the needed means for reconstruing possibility talk in terms of statements about the actual and their logical interrelationships. Accordingly, even if merely possible worlds are abandoned as a serious factor in ontology, it remains practicable to handle counterfactual discourse by other, significantly less problematic means.

Possible-world realism—also called modal realism—is the doctrine that, apart from this actual world of ours, the realm of being—of what there is—also includes alternative worlds that are not actual but merely possible. Being—reality at large—is something broader than mere actuality. There are two versions of the theory. Strong modal realism holds that those alternative worlds really exist, albeit in a separate domain of their own, outside the range of our universe's spacetime framework. And weak modal realism holds that although those alternative worlds do not really exist, they nevertheless somehow quasi-exist or subsist on their own in total disconnection from anything going on in this actual world of ours—apart, perhaps, of being thought about by real people.

Let us begin with the former. The most emphatic sort of strong modal realism proposed in recent times is that of David Lewis.[6] As he sees it, nonactual possible worlds are comparable to "remote planets, except most of them are much bigger than mere planets and they are not remote [since they are not located in our spatiotemporal realm]."[7] All of these worlds—and their contents—*exist* in the generic sense of the term, and all of them stand on exactly the same footing in this regard, although none exists in or interacts with another. (Existence in a narrower sense is always world-correlative, a matter of placement within some possible world.) This world of ours is nowise privileged in relation to the rest; it does not differ in ontological status but only in the parochial respect that we ourselves happen to belong to it. As Lewis puts it:

Our actual world is only one world among others. We call it alone actual not because it differs in kind from all the rest but because it is the world we inhabit. The inhabitants of other worlds may truly call their own worlds actual, if they mean by "actual" what we do; for the meaning we give to "actual" is such that it refers at any world *i* to that world *i* itself. "Actual" is idexical, like "I" or "here," or "now": it depends for its reference on the circumstances of utterance, which is the world where the utterance is located.[8]

As Lewis's approach sees it, the manifold of possible worlds as a whole is the fundamental ontological given. To be sure, we have no way to get

from here to an "elsewhere"-located realm of alternative possibility—it is
not that some other region of space is at issue. On this perspective, then,
there are no "unrealized possible worlds" at all—all possible worlds are
realized, all of them exist as parts of one all-comprehensive reality, one
vast existential manifold. It is just that they are spatiotemporally and
causally disconnected, there being no way save that of thought alone to
effect a transit from one to another. What Lewis in effect does is abolish
the idea of "nonexistent possibility" in favor of one vast existential realm
in which our spatiotemporal real world is only one sector among many.
(His theory is much like that of the Greek atomists, except that their
worlds were placed within a single overarching space and could collide
with one another.)

With Lewis, as with Spinoza, reality is accordingly an all-inclusive
existential manifold that encompasses the whole of possibility. He holds
that it is a fallacy "to think of the totality of all possible worlds as if
it were one grand world" because this invites the misstep of "thinking
that there are other ways that grand worlds might have been." Of course
the manifold of possibility could not possibly be different since whatever is possible at all is part of it. But this clearly does not block the
path to thinking of the totality of all possible worlds as one all-embracing
superworld.[9]

Lewis thus projects an (extremely generous) conception of existence
according to which: (1) anything whatsoever that is logically possible is
realized in some possible world; (2) the entire manifold of "nonexistent
possible worlds" is actually real; (3) the existential status of all of these
possible worlds is exactly alike, and indeed is exactly the same as our
own "real" world; and (4) there is also a narrower, more parochial sense
of existence/reality which comes to coexistence with ourselves in this particular world—that is, being colocated with ourselves in this particular
world's spatiotemporal framework.

But this position runs up against the decisive fact that one must "begin
from where one is," and *we* are placed within this actual world of ours.
There is no physical access to other possible worlds from this one. *For
us*, other possible worlds cannot but remain mere intellectual projections,
mere "figments of the imagination." The problem with Lewis's strong
actualism is that from our own starting point in the realm of the real—
the only one that we ourselves can possibly occupy—this whole business of otherwise existence is entirely speculative because our own access
to the wider realm beyond our parochial reality is limited to the route
of supposition and hypothesis.[10] Our standpoint is the only one in

whose terms our own considerations can proceed. The priority of the actuality in any discussion *of ours* is inevitable: it is not a matter of overcoming some capriciously adopted and in principle alterable point of departure.

But what of a weaker possible-world realism, one which, while holding that such worlds do not exist, nevertheless concedes them an attenuated form of being or reality in an actually detected domain of their own? Many philosophers deem even this sort of thing deeply problematic. As they were coming into increasing popularity, J. L. Mackie wrote that "talk of possible worlds ... cries out for further analysis. There *are* no possible worlds except the actual one; so what are we up to when we talk about them?"[11] And Larry Powers quipped that "The whole idea of possible worlds (perhaps laid out in space like raisins in a pudding) seems ludicrous."[12]

However, while this disinclination is plausible enough, the principal reason for rejecting the subsistence or quasi-reality of possible worlds lies in their cognitive inaccessibility. As the previous discussion has stressed, there is simply no viable way of identifying such merely possible worlds and their merely possible constituents. The problem lies in thinking that the locution "a world just like ours except that ..." can be implemented meaningfully. It cannot: as Burley's Principle indicates, once one specifies any change in the world's actual state of things, that "except that" listing can never be brought to an end. When we start to play fast and loose with the features of the world we cannot tell with any assurance how to proceed. Consider its law structure, for example. If electromagnetic radiation propagated at the speed of sound how would we have to readjust cosmology? Heaven only knows! To some extent we can conjecture about what consequences would possibly or probably follow from a reality-abrogating supposition. (If the law of gravitation were an inverse cube law, their significantly lesser weight would permit the evolution of larger dinosaurs.) But we cannot go very far here. We could not redesign the entire world—too many issues would always be left unresolved. In a well-articulated system of geometry, the axioms are independent—each can be changed without affecting the rest. But we have little if any knowledge about the interdependency of natural laws, and if we adopt a hypothesis to change one of them we cannot concretely determine what impact this will have on the rest. The specification of alternative possible worlds is an utterly impossible task for us finite mortals. Even when viewed epistemically as mere methodological thought-tools, merely possible worlds and individuals remain totally impracticable.

3   Averting the Problem of Transworld Reidentification

But one might object that we achieve merely possible individuals via such scenarios as:

He thought he saw a man in the garden; and *that man* wore a top hat and carried a cane.

She thought there was a ghost on the landing. And *this ghost* wore a long white robe and made strange wailings.

Nevertheless, in such cases we do not properly speaking have an anaphorical reference (*that man* or *this ghost*) to a "nonexistent object"—to something that does not in fact exist.

To see this it is necessary to note two facts. First, letting Tp represent that someone (he, she, they, whatever) *thinks p* to be the case, observe that:

(A) $T(\exists x)Fx$ (thinks that there is an x that has F)

is something very different from the claim:

(B) $(\exists x)TFx$ (There is an x which Z thinks to have F)

The second thesis involves an existential claim on *our* part; the first leaves us entirely out of it in point of any existential claim or commitment. It is certainly (and unproblematically) possible for people to *think* about nonexistent individuals, but that neither entails nor requires that there actually be nonexistent individuals for them to think about. We must be very careful not to conflict or confuse theses (A) and (B).

The second important consideration is in that the two preceding statements actually come down to:

He thought he saw a man in the garden who wore a top hat and carried a cane.

She thought there was a ghost on the landing which wore a white robe and made strange wailings.

The appropriate grammar of these statements is not

$T(\exists!x)Fx \ \& \ G(\iota x \ni TFx)$

but rather

$T(\exists!x)(Fx \ \& \ Gx)$

With the former formulation we stake a claim that is made on our own account, namely $G(\iota x \ni TFx)$, which commits *us* to attaching G to an object of a certain sort to whose existence we stand committed. But this is nowise the case. The entire contention at issue is (and should be seen as being) within the scope of T. It is the subject at issue (Z) who bears the total responsibility for any purportings of existence; no existential commitment of any sort rubs off on us, who are merely the *reporters* of Z's odd belief in objects that do not exist.

After all, there is a significant difference between "I think that there is a possible world with the feature F" (symbolically $T(\exists w)Fw$) and "There is a possible world that I think to have the feature F" (symbolically $(\exists w)TFw$). For the former does not entail the existence of possible worlds: $T(\exists w)Fw$ can be perfectly true not only when both $(\exists w)TFw$ and $(\exists w)Fw$ are false, but even when *very* statement of the fact $(\exists w)(w \neq w^* \& [\ldots w \ldots])$ is false—w^* here being the *actual* world—because there just are no "merely possible worlds" at all. Neither here nor elsewhere does one bring something (except a thought) into existence by merely thinking it to exist.[13]

In the end, your statement in that just quoted passage is just inaccurate: you are not thinking of a (specific) world but one of a manifold encompassing *any* world that meets certain conditions. It is not just that some crucial facts of the matter about that purportedly concrete world are unknown but that they are *missing*. To reemphasize: we are once again dealing not with concrete worlds but with the generic scenarios of de dicto possibilistic discourse.

One positive result of rejecting possible individuals and worlds is that the vexed question of the "transworld identity" across the range of merely possible individuals will now never arise.[14] Lewis proposes to settle this issue of "counterparts," as he calls them, on the basis in similarity. But this clearly will not do, since with things as with worlds there is only similarity of respect: there is, after all, no such thing as synoptic similarity but only similarity in regard to this or that particular aspect.[15]

The most sensible view is that transworld identity is simply a matter of supposition (of assumption or postulation). There is no supposition-independent fact of the matter about it. Thus, consider what might be called Chisholm-type questions[16]: Two individuals, A and B, gradually interchange their properties (first their height, then their weight, then their hair color, etc.). At what point does A become B? Answer—it's all up to you. It's your hypothesis that's at issue here. You have to tell us how you want it to play out. The issue is once more one of questioner's prerogative. There are no supposition-external facts of the matter to constrain this. Once we acknowledge that assumption and hypothesis alone provide our pathway to the domain of the merely possible, this vexed question of the "transworld identity" of individuals becomes simplicity itself. Given that nonexistent worlds are constituted by hypotheses (by imaginative stipulations) the identity of their individuals is dependent entirely on the nature of the hypotheses at issue. Is that imaginary general identical with Caesar or not? Ask the individual whose imagination is at issue! It will all depend on what he says.

In a classic paper, W. V. O. Quine (1948) sought to embarrass the possible-individual theorists of his day with the question: "How many possible fat men are there in the doorway?" His opponents proceeded to seek refuge in possible worlds, attempting the response that there was one such individual per world, at best. But in taking this line, however, they failed to note that essentially the same difficulty with identifying possible fat men arose with identifying the worlds that supposedly contain them. Even when world w puts its single fat man into doorway no. 1, how many other doorways does it have for occupancy by thinner men?

The long and short of it is that in refusing to grant any sort of existence or reality to nonexistent possible individuals and worlds averts a plurality of intractable issues. Let us examine some of them.

4   Worldly Woes

Charles Chihara had proposed viewing the problem of nonactual possible worlds in an epistemological light, arguing that "it is not clear that we have any conceivable way to gain knowledge of other possible worlds" so that "knowledge of such worlds is completely beyond our powers."[17] And this is both sensible and correct as far as it goes. But the real problem is not just the limit of our cognitive powers with respect to other-worldly things but the inherent improbability of characterizing and identifying such a "world"—of specifying what is and what is not at issue with any particular instance of this sort of thing. The problem, in other words, is not with ourselves but with those "worlds." It is not just knowledge of such objects that is infeasible but the very "objects" themselves.

As William Lycan has observed, "a view according to which worlds are the way we say they are because they are simply stipulations by us has a considerable advantage over [a supposedly self-sufficient realm of nonexistent]."[18] But, of course, what we can actually ever say about worlds (*in abstracto*) never suffices to identify *a* world. When possible-world theorists propose to identify worlds via sets of propositions, they fail to recognize that actually available (and thereby finite) sets of propositions simply cannot do the requisite job. For anything worthy of being designated as a world will have to involve a plethora of descriptive detail that can never actually be articulated.

Robert Stalnaker has written:

> There is no mystery to the fact that I can partially define a possible world in such a way that I am ignorant of some of the determinate truths in that world. One way I can do this is to attribute to it features of the actual world which are unknown to me. Thus I can say, "I am thinking of a possible world in which the population of China is just the same, on each day, as it is in the actual world." I am making up this world—it is a pure product of my intentions—but there are already things true in it which I shall never know.[19]

One must, however, be careful here. Those "incomplete worlds" are not worlds at all, so that in effect no one single definitely identifiable world is in view but at best entire spectra or families thereof. It is *not* the case that some one individuated world is before us, only one that is in some respects inherently incomprehensible, so that there are facets about *it* that cannot be known. There simply is no definite "it" about which there are certain facts of the matter that we cannot determine.

Consider the Stalnaker-analogous supposition of a world where the population of Shanghai is otherwise just exactly as is, except that oneself is also among them. Can such a world count as other than schematic? Well, perhaps, so as far as its *people* go. But what about the rest of its makeup? How did I get there? Where are those air particles that my body displaces? And how did they get there? What that assumption has done is to confront us once again with something that is not a world but a world schema. And the fact remains that the reason for ignorance about matters of content is simply that the world to be at issue has been characterized only partially, so that in effect no one single definitely identifiable world is in view.

It might seem possible to arrive at possible worlds via fictionalizing assumptions, as per: "Assume a possible world in which dogs have horns." But this, of course, takes us no further than certain radically incomplete "states of affairs" and does not put any concrete particulars on the agenda. In effect it deals not with authentic worlds but with a schematic supposition of propositional possibilities. Thus consider: "Assume a world whose dogs have horns." How are we to manage it? How did those dogs come by horns—by sorcery? By cross-breeding with goats? By a different course of evolution? The reality of things will always go beyond what we are explicitly instructed to assume. Assumptions can never suffice to characterize an authentic world. The descriptive specification of a whole world—any world—is an impossible task. The assumptions that supposedly take us into nonexistent possible worlds are always incomplete and thereby merely schematic.[20]

A merely possible individual or world that is only "partially defined" by way of an assumption or supposition is in effect a *schema* to which a plurality of definite possible individuals (or worlds) can in principle answer, exactly as a partially described actual individual can (for all we know) turn out to be any one of a plurality of alternatives. Speaking of "partially defined individuals"—an expression favored by some theorists—makes about as much sense as speaking of a "partially pregnant woman." Such incomplete specifications of individuals confront us with possible individual world *schemata* rather than *individuated* possible individuals or worlds, and indicate an indeterminacy that is epistemic rather than ontological. A striking fact about schematically identified individuals is that the Law of Excluded Middle—in the form of the principle that of thesis p and its contradictory not-p, one must be true—fails to obtain. Thus, in the context of the hypothesis "Assume there were a red-headed person sitting in that chair," we could neither say that this hypothetical person is male nor that it is not, nor again, neither that it is six feet tall nor that it is not. Indeed this very failure, even in principle, of a complete characterization precludes the prospect that we are dealing with an authentic *world*.

To be sure, some theorists, while rejecting the existence/subsistence of nonactual possible worlds as such, are prepared to endorse the existence of certain abstract entities that qualify as "world descriptions" or "maximal state-of-affairs characterizations."[22] But this does no more than to shift the problem to another mode of impracticability owing to the infeasibility of ever arriving at such an abstract entity—a viably world-characterizing description. The task of providing something that would be qualified as such is simply unachievable. Possible-world theorists cannot produce a single adequate example of the kind of thing they so glibly purport to deal with.

Some semanticists cover the inescapable indefiniteness of possible individuals and worlds by treating them as abstract objects of a particular sort.[23] But this is deeply problematic; abstractions cannot achieve concreteness, and individuals and worlds must be concrete. Although abstractions may *characterize a type of* world—and lots of them—they do not *identify a single one* of them—any more than an abstraction like "bluish" determines any particular blue color patch to which (among others) it appertains.

Thus, possible-world theorists are caught up in a position-destructive dilemma. They can either hold that the merely possible individuals that populate such a world are just like real-world ones in point of descriptive definiteness—in which case they have no practicable way of identifying or individuating such particulars, and thereby no practicable way of building up their possible worlds. Or else they can be content with descriptive individuation of a practicable kind at the level of schematic generality—thus indeed achieving a meaningful basis for discussing possible worlds—but then they have to deal at a level of generality with "worlds" whose nature is schematic and whose individuation is unachieved and unachievable. They cannot have it both ways.

5   Redistribution Worlds Are No Exception

"But surely it's not so difficult to specify worlds. A variant world could, for example, be just like this one except for Caesar's deciding not to cross the Rubicon on that fateful occasion." Very well. But now just exactly what is this world to be like? What actually happens in such a world? An endless proliferation of questions arises here. Does Caesar change his mind again a moment later and proceed as before with just a minor delay? Is he impelled across by *force majeure* and then decides to carry on as he was? And if he doesn't cross, then exactly what does he do? And what will all those who interacted with him then and later be doing instead? The resulting list of questions is literally endless.[24] Innumerable alternatives confront us at every point and these themselves lead to further alternatives. As we specify even more detail we do not reach definite worlds but continually open doorways to find further possibilities. The idea of identifying a possible world in some descriptive way or other is simply unworkable. (To be sure, even this actual world is not adequately describable by us, but that does not matter for its identification, seeing that, most fortunately, we can—thanks to our existence within "this actual world of ours"—individuate its contents ostensively.)

Again, consider the hypothesis: "Suppose a world otherwise just like our actual one except that there is an elephant in yonder corner of the room." Contrary to first appearance, this supposition also introduces not a particular (individuated) world, but rather a world-schema that can be filled out in many alternative ways. Thus consider the questions: Are we to redistribute the actually existing elephants and put one of them into the corner (if so—which one)? Are we to take some actual thing—say the chair in the corner—and transmute it into an elephant (assuming that such a supposition could qualify as feasible)? Are we to keep all the actual things of our world's inventory in existence and somehow "make room" for an additional, supernumerary one—the (hypothetical) elephant at issue? Until issues of this sort have been resolved, the supposition does not introduce a definitely specified world into the framework of discussion— any more than the supposition "Assume there were a red-headed person sitting in that chair" succeeds in introducing a definitely identified person.

But even if it is not practicable to realize individual nonexistent possible worlds through populating them with nonexistent objects, could we not at least create them by redeploying existing objects?[25]

Wittgenstein's *Tractatus* proposed what William Lycan called a "combinationalist" construal of other possible worlds via the idea that such worlds are simply rearrangements of what in actual fact are the ultimate constituents of this world.[26] As David Lewis rightly emphasized, any recombinant theory of possibility will have to make up its mind as to the nature of the basic building blocks.[27] It cannot avoid the question: Recombinations of what? And here the two prime prospects are: (i) recombinations of physical objects (perhaps at the atomic or even subatomic levels) in the framework of space and time, or (ii) recombinations of some of the descriptive properties of otherwise invariant objects. Either way, we take a fundamentally conservative view of possible worlds: all of them contain the same basic objects as their actual one (no additions or deletions allowed!) but with changes only in either (i) location in spacetime, or (ii) possession of certain (inessential) descriptive features. Such recombinant possible worlds can either shuffle actual objects about in spacetime or shuffle their properties about a descriptive "space."[28] Let us consider these two prospects.

Consider the idea of obtaining a world via a hypothetical *spatial rearrangement* of actual objects. So, for example, we might populate *that* shelf (pointing) with *these* particular books (pointing) that are currently placed on other shelves in the bookcase. Or again we might proceed by hypothetically stipulating that yon two cats, Tom and Jerry, be interchanged on their respective mats. Then would not the hypothesis "Let us suppose that Tom were on mat 2 (instead of mat 1 as is) and Jerry were on mat 1 (instead of mat 2 as is)" succeed in projecting another possible world for us? Alas, no. For consider once more how we could ever manage to get there from here. Assume those two cats to be interchanged *right now*. Where were they a nanosecond ago? In their actual places—and if not, then where? And how did they effect their transit? How are the laws of physics to be changed to accommodate these changes of place? Or if we hypothetically shuffled the books in that bookcase about, then what are we to do with the course of world history that has placed them in their present positions and will subsequently position them in their future locations?

Problems of laws of nature apart, another difficulty with these positional redistribution worlds is that they rest on the naive doctrine of ancient Greek atomism that space is something entirely independent of the objects that "occupy" it, so that a uniform space is available for objects to be redistributed in it. Given a realistic physics that involves space and its objects along with time and its processes, these redistribution worlds are no longer available as such.

The trouble with spatial rearrangement worlds is that they open up a vast host of questions as to how to get there from here to which there just is no available answer. The hypotheses themselves do not identify a concrete particular. They too only result in something schematic; the necessary concreteness is not achieved, because too much remains unresolved: the exact positing of the constituents, the exact processes by which they came to be where they are, and so on. In hypothetically exchanging or interchanging those various real-world objects we emphatically do not effect a transit into another particular possible world, but simply put on the agenda a vast spectrum of alternative possibilities that would, for individuation, require a filling in of a volume of detail which we simply cannot provide in the necessary completeness. Such hypotheses lead down a path where every answer to a question that arises leads to yet further questions, *ad infinitum*.[29]

A different sort of rearrangement world has also been contemplated—one in which we are to rearrange not *positions* but *properties*. "Suppose a possible world just like this one except that its elephants are pink." However, such a supposition is literally absurd. We cannot possibly implement its indication of "just like." For we cannot get those elephants to be a different color without changing the visual apparatus of vision-capable creatures—or, even more drastically, the laws of nature that are operative in such a world. And this sort of thing is clearly bound to have wider ramifications; when we change visual resources or natural laws, the sort of world that results will obviously not be "just like" this one except in point of elephant coloration. But would one not obtain a particular world by changing the aforementioned specification to read "except for its elephants being pink and whatever else needs to be changed to realize this end"? Alas, no. For this revision once again does not lead to any single world whatsoever, but a vast manifold of them. There will inevitably be many different possible routes to effect that specified result, and the absence of uniqueness is again fatal to the prospect of world specification. Those property redistribution hypotheses themselves accomplish no more than to project ever-schematic scenarios. Those descriptively recombinant "worlds" are an illusion.

But what of the prospect of rearranging not objects or properties but truth values? Can we not simply coordinate possible worlds with specifications that assign the truth values T or F to every sentence of a language? Not really! For that assignment would have to be carried out in a logically coherent way (e.g., we cannot assign T to p and T to q and F to $p \& q$). And there is no effective way in which this can be accomplished for an entire language on a global basis except in the special case when the actual world is being characterized. (This is again the lesson of Burley's Principle.) And given the logical and metaphysical integrity of the real, it could not be otherwise. As Burley's Principle shows, once we start conjuring with the real world's facts we are on a slippery slope to nowhere. Rearrangement proposals are simply unable to overcome the underdetermination that affects fact-violating suppositions in general.

Possible-world semanticists talk and reason as though possible worlds were somehow given, were part of what is available and in hand to work with. Where these worlds are to come from—how we can get there from here—is a question they simply ignore. They never tell us how we are to arrive at possible worlds given our *de facto* starting point in this one. They proceed as though one could obtain by mere fiat that which would have to be the work of honest toil—albeit a labor which, however extensive, is bound to be unavailing in the present case.

6   Counterfactuals and Possible Worlds

As we have seen above, even though counterfactuals look to unrealized circumstances, they are always projected from a vantage point within this actual world and are predicated on its situations and concepts. They answer *our* what-if questions: What if wishes were horses? What if pigs could fly? What if there were a prime number between 3 and 5? Such questions project suppositions and are predicated upon them. And they eventually are within a larger context of environing beliefs—beliefs that we in fact hold.

The tenability of counterfactual contentions hinges entirely on issues of precedence and priority among beliefs—on matters of priority and precedence in case of potential conflict. And it deserves stress that these propositional priorities are dependent on how things stand *in this world*. The idea that counterfactuals relate to "other possible worlds" rests on very questionable footing. It is certainly not true (in this world) that Napoleon was born in New York. But a counterfactual like "If Napoleon had been born in New York, then he would be an American" is not a fact about other possible worlds in which Napoleon was born in New York. Its foundation is a fact about this world as much as any—the contingently true and this-world-characterizing circumstance that in view of the politico-geographic location of that city, people born in New York are Americans. The various examples considered above show how such a mechanism of priority in fact operates. (Nonexistent possible objects— let alone possible worlds—do not come into it.)

Contemporary philosophers incline almost without exception to the view that alternative possibilities require alternative worlds. Thus consider "If there were no cats in the world, dogs would be even more popular as pets than they now are." Regarding this counterfactual, these theorists would maintain that we cannot really make sense of it without contemplating possible worlds, alternative to ours, from which cats are missing. But the presently contemplated approach to counterfactuals puts this view into a very dubious lights. For consider the situation specified via the following beliefs:

(1) Some people who do not own dogs would get one if cats were not available.

(2) Cats are available.

(3) There are n dog owners in the world (for a suitable value of n).

And let it be that we are instructed to suppose (2) to be false.

The assumption explicitly requires us to replace (2) with not-(2): "Cats are not available." In the face of this supposition, we must abandon either (1) or (3) if consistency is to be maintained. For (1) and not-(2) combine to yield the conclusion that the number of dog-owners would *increase* beyond its present size, and this is incompatible with (3). But when—as usual—we prioritize general tendencies over specific facts, we shall arrive at the counterfactual in question. No recourse to nonexistent worlds is required to make good sense of this counterfactual. It is quite enough to consider real-world actualities and the relations of epistemic priority or precedence that obtain among them.

It is necessary in this context to distinguish between *world-altering* counterfactuals and *world-replacing* ones. The former takes the generic format: "If the world were *F*, then. . . ." Here we have such instances as: "If large mammalian species had evolved much earlier, then fewer organic species would exist today" or "If the big bang had resulted in a hotter plasma, stable chemical elements would not have come into being," or "If the law of gravitation were an inverse cube law so that gravitation were a much weaker force then there would be far fewer planetary systems with potentially life-supporting planets." Such hypothetical modification of the actual world leave enough in place that coherent inferences can be drawn from them.

However, the situation is very different with world-replacing hypotheticals of the form: "If a world entirely different from this one existed in its place, then. . . ." In the wake of such a hypothesis—one which replaces this world with something altogether different—no coherent conclusion can be drawn. Once we envision that radical a change all cognitive bets are simply off; the rug is pulled out from under our feet, and we are now at sea without any compass to give us an idea which way to turn. That radical world-abrogating hypothesis leaves us adrift on a sea of incomprehension. Any prospect of drawing a meaningful conclusion is denied us. The challenge of specifying an *alteration in* the actual world is producible for us, but that of specifying a *replacement for* the world—a genuinely alternative possible world—is beyond our powers. The conjunction is "Keep as is only what you are explicitly instructed to maintain and consider everything else as open to change." And here we just don't have a clue as to how to carry this instruction though. Changing *everything*—or even merely everything else—is simply too much for us. In this sort of situation the resources required for a meaningful process of counterfactual reasoning are denied to us.

The long and short of it is that although we cannot come to satisfactory cognitive grips with "nonexistent possible *worlds*," we can indeed reason quite well from more concretely limited counterfactual suppositions based on nonactualized possibilities. But in reasoning here we need not look outside this world of ours to other possible worlds, but need only look inside to the epistemic priority status of our cognitive commitments. What we are called on to do is simply make a comparative determination of post-assumption retainability among real-world truths. There is no need, fortunately, to enter into the problematic contemplation of the actuality of other, nonexistent worlds of which—if the preceding arguments hold good—we cannot make proper sense. A recourse to the epistemic priorities that govern our commitments regarding our claims about the world is quite sufficient to deal satisfactorily with counterfactuals.

Model semanticists appear to think that they are clarifying matters in going from "obscure" counterfactuals to "unproblematic" nonexistent worlds. They could not be more wrong: to analyze counterfactuals via nonexistent worlds is to explain what is obscure by what is yet more so.

Fortunately, then, counterfactuals need not carry the crushing weight of possible worlds on their back. As we have seen, the crux of counterfactual analysis is a matter not of scrutinizing the situation in other possible worlds, but at most one of prioritizing our accepted beliefs regarding this actual world. There is simply no need to look beyond the cluster of the environing propositions that are immediately relevant to the particular counterfactual at issue. We certainly have no need to become involved in anything as grandiose as a possible world: dealing with a very moderate sector of reality suffices for all our needs. Counterfactual reasoning is a cognitively crucial device that neither demands nor invites an outlandish ontology of other possible worlds.

And in particular, causal counterfactuals on the order of "If Susan had dropped the vase it would have broken" are obviously facts about the real world—it's just that they are facts about its *modus operandi* rather than about its occurrences and events.[30]

And not only are possible worlds not necessary for construing counterfactuals; they are not even sufficient. It is a decisive drawback and deficit of the possible-worlds approach to counterfactuals that it is inherently impracticable in various otherwise unproblematic cases. Thus consider the counterfactual: "If four were greater than five, then arithmetic would be involved in a contradiction." We clearly cannot handle this by contemplating the situation in those possible worlds where four is greater than five, since there obviously are not any. Nevertheless, no one would have any difficulty making sense of that counterfactual, and in fact the present aporetic analysis validates it straightforwardly. In the setting of our present approach, however, matters run smoothly. We have the following accepted propositions:

(1) Four is *not* greater than five.

(2) The consistency of arithmetic [as we know it] entails (1).

(3) Arithmetic is—and ought to be—consistent.

And now, with not-(1) assumed, we are at once forced into an abandonment of either (2) or (3), seeing that the trio {not-(1), (2), (3)} is logically
inconsistent. Since there is no viable way around (2), this means that we
will have to give up (3) and see arithmetic as involved in contradiction.
And this validates the counterfactual under consideration. A possible-
worlds analysis, by contrast, could get nowhere here.[31]

Again, consider: "If there were twice as many integers between two and
five as there in fact are, then there would be five integers between two and
six." This too is smoothly amenable to our world-dispensing analysis. We
have as our accepted theses:

(1)   There are two integers between two and five (namely three and four).

(2)   There is one more integer between two and six than between two
and five, namely five itself.

And we are asked to assume there to be twice as many integers between
two and five as there actually are (viz. twice two, as per (1)). The resolution here is straightforward. We must drop (1) as per the assumption.
Moreover, we obtain four integers between 2 and 5—again as per the
assumption. And between 2 and 6 we obtain one more as per (2). This
makes five in all, just as the counterfactual states.

Such impossibilistic counterfactuals create hopeless difficulties for a
possible-world analysis. But the belief-repair analysis discussed in the
present book can accommodate them without difficulty. William Lycan
has rightly observed that "we 'understand' counterfactuals in ordinary
conversation, but for purposes of serious philosophy they have proved to
be among the most troublesome and elusive expressions there are."[32]
However, this persumptive view seems to be the case only because recent
philosophers have insisted on making matters difficult for themselves by
bringing possible worlds into it. The fact is that the analysis of counterfactuals is at most a matter of a very localized propositional prioritization
in line with some rather straightforward and pragmatically cogent rules.
Given such right-of-way prioritization, the mystery of troublesome elu-
siveness is completely dissipated. And vast manifolds such as worlds do
not come into it at all. As with an Agatha Christie detective story, a
closer scrutiny of the proximate suspects immediately involved will suffice
to resolve the mystery.

As the preceding discussion has indicated time and again, the acceptability of counterfactuals is a matter of our convictions regarding this-
worldly arrangements, not world-abrogating conjectures. All that is ever

required for the analysis of counterfactuals is a handful of plausible ground rules of precedence and priority to settle matters of conflict resolution—of "right of way" in a conflict of aporetic inconsistency among the immediately involved reality-appertaining propositions. And operating in this way does not call for any *global* device of the sort at issue with possible worlds; all we ever need is a *local* device for assessing the comparative priority of a few propositions.

Establishing a counterfactual need not take us outside the familiar and comfortable region of our own beliefs and does not require us to exit from the realm of reality to concern ourselves with other, unrealized possible worlds. What is at issue is a localized micro-process and not a globalized macro-process: belief-contravening suppositions do not shift the frame of reference to other possible worlds but merely test the comparative solidity and staying power of our actual claims in immediate contextual neighborhood. To have recourse to a metrically structured manifold of possible worlds to settle the questions of the assertibility of counterfactuals would be like using a sledge hammer to squash a gnat. The pivotal lesson is thus clear: to validate counterfactuals as informatively productive assertions we need not go so far as to launch conjectural global forays into other-worldly domains. It suffices to take note of the relevant local ground rules about the prioritization of claims in the setting of this actual world.

Granted, the antecedents of belief-contradicting conditionals seemingly carry us into the realm of alternative reality—of fiction. In making the belief-contradicting assumption that the antecedent of such a conditional involves, we adopt a scenario that carries us outside the confines of the real. However, assessing the tenability of the conditional fortunately does *not* require any excursions into other fictional, unrealized worlds. We simply have to probe the priority structure of the manifold of our immediately relevant beliefs. Analyzing such conditionals never requires more than making consistency-restorative deletions from among our beliefs regarding the real world. All that is ever needed for assessing the tenability of counterfactuals is to determine which of our reality-characterizing beliefs must give way to which others in the particular context of deliberation.

The pivotal fact is that counterfactuals always rest on factual foundations. Thus the factual thesis

The vice president always presides at cabinet meetings when the president is absent provides the basis for the factual conditional

Since the president is absent, the vice president presides,

as well as for the (agnostic) conditional

If the president is absent, the vice president presides.

But it also provides the basis for the counterfactual

If the president were absent (which he's not), the vice president would preside (which he does not).

And so, counterfactuality not a matter of leaping from the real world into an murky manifold of unrealized possibility, but simply requires a duly adjusted view of what we believe about this actual world. On the present approach the validation of counterfactuals is always merely a matter of making deletions from among our beliefs in a way that (1) restores overall consistency in the facts of that counterfactual hypothesis, and (2) does this by way of abandoning those salient beliefs that have the lowest priority standing in the relevant context of deliberation. It is thus quite enough to look to real-world actualities and the relations of epistemic priority or precedence that obtain among them.

Yet consider the following objection: "You say that counterfactuals tell us about the world. At the same time you proceed on the basis not of the *truth* conditions of counterfactuals but conditions for their *assertibility* or *acceptability*. Is there not a discrepancy here?" Not really: no significant discord arises. After all, we can only address the question of what the world in fact *is* like via what we consider appropriate for us to say about it. Our only route to information about the world is via the epistemic issue of what one can reasonably and justifiably claim. We have no alternative way to answer the question "What *is* the world like?" other than via what we find appropriate to accept or assert about it. "Tell me what the world is like as apart from what you think it to be" is an instruction one cannot obey.

Back then to the main point. The treatment of subjunctive and specifically counterfactual propositions by contemporary possible-world theorists is both inherently problematic and optimistically complicated. The fanciful complexities they introduce into their analyses is such as to make one wonder how it can be if *that* is the way to understand these conditionals that communication about these matters in ordinary language can proceed in the comparatively sure-footed and unproblematic way in which it ordinarily operates. Fortunately, no such foray into belief-inconsistent worlds is required for a doxastic approach, which only calls for some plausible prioritization among saliently relevant beliefs. Here no fanciful complexities about possible worlds accessible from other possible worlds manage to arise. All that is required is the application of a perspicuous prioritization rule to a handful of saliently relevant beliefs. We need never to look further than the situation of this real world of ours, deploying real-world oriented considerations that function in a natural way to give epistemic priority to some factual contentions over others.

Conclusion

In closing, it is useful to summarize the principal conclusions emerging from these deliberations regarding conditionals.

The first is that all types of conditionals can be understood and accounted for in terms of logico-conceptual derivability, so that deductive inference ($\vdash$) constitutes the basis of conditionality ($\Rightarrow$) in general.

The second is that when implication ($\rightarrow$) is construed as embracing specifically and only those modes of conditionality that obey certain standard logical principles (*modus ponens*, transitivity, monotonicity, contrapositivity, and conjunctivity), then not every mode of conditionality represents an implication relationship. And, in particular, counterfactual conditionalization will certainly not do so.

The third lesson is that *all* of the standardly considered modes of implication ($\rightarrow$) can be explicated uniformly in terms of contextual implication ($[S]\!\!\mapsto$) as specified by the definition:

$$p \; [S]\!\!\mapsto q \text{ for } (p + S) \vdash q$$

with respect to suitable propositional sets S. This transpires as follows:

1. With *deducibility* itself ($\vdash$) we have $S = \varnothing$ (the null set of propositions).

2. With *material implication* ($\supset$) we have $S = T$ (the set of truths).

3. With *strict implication* ($\prec$) we have $S = N$ (the set of necessary truths).

4. With *doxastic implication* ($[B]\!\!\mapsto$) we have $S = B$ (a set of accepted beliefs).

This points the way to the fourth key lesson, namely that the most commonly encountered sort of conditionals—the doxastic—are not always to be classified under the duality of true/false, but should properly be positioned in the spectrum of the more or less plausible. And the plausibility of such a conditional is exactly that of the least plausible component of the most plausible belief-afforded means for closing the inferential gap from antecedent to consequent. (This is the "minimax" principle of plausibility determination based on weakest-link considerations.)

The fifth lesson is that whereas factual conditionals can be handled in terms of deducibility from our belief-commitments, the analysis of counterfactuals will require some rather demanding restrictions of the range of our case-relevant beliefs in the interest of maintaining coherence and consistency. Counterfactuals hinge not on the manifold B of our belief-commitments at large, but rather on a very limited group of specifically encountered beliefs **B** that are relevantly salient to the hypothetical "what-if" question at issue.

The sixth lesson is that with counterfactuals the crucial factor is that of knowing certain standard groundrules for *priority* inherent in the economy of information management. Specifically, the prioritization of generality and lawfulness becomes especially prominent.

The seventh key lesson is that various modes of inferential reasoning have prioritization ground rules for belief-retention that differ in characteristic ways geared to the purposive context at issue. Table C.1 depicts some of the relationships that obtain here.

The eighth key lesson is that once the ⊢-relationship of logico-conceptual deducibility is in hand, the explanatory analysis of all conditional relationships—counterfactual conditionalization specifically included—requires us to look no further than the realm of truth (or what we accept as such). And specifically, in no case does the validation need to invoke merely possible worlds with their nonactual states of affairs and unrealized individuals. For the tenability of any particular counterfactual can be assessed simply on the basis of our views regarding the actual world. The epistemologically determinate issue of the substance

Table C.1
Contextual Priority Variation

Mode of Reasoning	Priority Ranking	
	Priority of the pivotal hypothesis	Prioritizing of the accepted beliefs
1. *Ordinary hypothetical reasoning*	top priority	not applicable (no possible conflict)
2. *Inductive reasoning*	top priority	in reverse order of informativeness
3. *Counterfactual reasoning*	top priority	in order of informativeness
4. *Reductio ad absurdum*	bottom priority	all alike have top-priority
5. *Per impossibile*	top priority	in order of systemic informativeness

and nature of our beliefs about the real suffices entirely to meet the needs of the situation, with no need for recourse to a suppositionally projected realm of unrealized possibility. An epistemic approach of counterfactuals thus provides a comparatively simple and natural means to their appropriate analysis.

To be sure—and this is lesson number nine—counterfactual analysis has ground rules that transcend the realm of strictly factual, purely truth-oriented deliberation in invoking essentially epistemic considerations both of issue-salience and also of informativeness as a guide to precedence and priority in matters of conflict resolution.

Finally, a tenth lesson is that the acceptability-assessment of counterfactuals is best handled not on the basis of semantic (truth-orientated) relationships but rather via epistemic considerations. In the end it is not just the content but also the epistemic architecture of the manifold of our beliefs that determines the appropriateness of counterfactual conditionals.

Notes

Chapter 1 Fundamentals (Pages 1–14)

1. Some writers require thought experimentation to call for a supposition that is known or believed to be false. But this is in fact only one, particularly strong form of this process. When the detective reasons, "Now suppose that the butler did it …" at some early stage of the investigation, his reasoning clearly qualifies as a thought experiment even should it possibly and perhaps even probably turn out in due course that the butler indeed did it.

2. "Ueber Gedankenexperimente," in Mach 1905, pp. 183–200 (see p. 187).

3. For an interesting discussion of scientific thought experiments see Kuhn 1981.

4. On these issues, see Lycan 2001.

5. For an elaborate discussion of this thesis see Dummett 1981, pp. 348–352.

6. Lycan 2001 considers more subtle aspects of this issue.

7. On this issue see Anvera, "Conditional Perfection," in Athanasiadou and Dirven 1997, pp. 169–190.

8. On "as-if," see Adams 1993, p. 5.

9. See Bernard Conree in Traugott et al. 1986, p. 79.

10. Ibid.

11. See Traugott's essay in Athanasiadou and Dirven 1997, pp. 165–167.

12. For an elaborate discussion of "The Use of Conditionals in Inducements and Deterrents" see Fillenbaum in Traugott et al. 1986, pp. 179–196.

Chapter 2 Matters of Aspect (Pages 15–30)

1. Many languages differentiate between such "since" and "if-then" conditionals—not Indo-European languages alone.

2. An instructive discussion of linguistic issues regarding the temporal aspects of conditionals is given by Conree in Traugott et al. 1986, pp. 93–96.

3. Note that the conditional "If he has overeaten, then his stomach will usually ache" has to be contraposed into something like "If his stomach does not ache, then he has probably not overeaten."

4. On these issues, see Dale in Athanasiado and Dirven 1997, pp. 97–114, as well as Tynan and Lewin, in the same volume, pp. 115–142.

5. Medieval logicians already recognized the importance of this distinction between the modification of a *consequence* and one of a *consequent*. (See note 6 below.)

6. Recognition of this distinction goes back (at least) to Boethius' discussion of divine fore-knowledge in Book V of *The Consolations*. See Stewart, Rand, and Teolos 1973.

Chapter 3   Modes of Implication (Pages 31–50)

1. See Rescher 1968, Haak 1978, Gable 2000, and Jacquette 2002.

2. Mackie (1973a) calls this the "condensed argument account" of conditionals. It goes back to F. P. Ramsey's studies of the 1920s, but only gradually found exponents later on.

3. Bradley 1922, p. 86.

4. One of the reasons for imposing various of these requirements upon implication lies in the consideration that strict implication ($\prec$) is intermediate in strength between material implication ($\supset$) and entailment ($\vdash$)—seeing that $p \vdash q$ yields $p \prec q$ and this in turn yields $p \supset q$. If implication is to cover the whole of the intervening spectrum, then it seems plausible to stipulate that any generalization that holds *both* for $\vdash$ and for $\supset$ should also hold for $\rightarrow$.

5. Hilbert and Ackermann 1950.

6. For example, in Ziehen 1920, pp. 394, 702.

7. Actually the "beliefs" at issue are not *occurrent beliefs* but *belief commitments*, seeing that B is supposed to be closed under derivability ($\vdash$).

8. Kneale and Kneale 1985, p. 134.

9. Strawson 1952, p. 85.

10. Hacking 1998, p. 3.

11. On cases of this sort, see Dale 1972.

12. On quasi-truth-functionality, see Rescher 1962.

13. Quoted in Sanford 1989, p. 64 (n. 1). See also Grice 1989.

14. For further details and references to the literature, see Rescher 2001.

15. Sextus Empiricus, *Outlines of Pyrrhonism*, bk. II, secs. 110–112.

16. Ibid.

17. And so, since $\vdash P$ entails $\Box P$, all Chrysippean implications are strict.

Chapter 4   Conditional Complications (Pages 51–58)

1. Dummett 1973, p. 349.

2. The theory of monotonic inference is discussed at length in Levi 1996, especially chapters 5–7. Failure of monotonicity is general to nondeductive inference and has nothing to do with the specific issue of counterfactuality.

3. Compare section 1 of chapter 1.

4. Cited in Sanford 1989, p. 93.

Chapter 5   Doxastic Implication and Plausibility (Pages 59–66)

1. One's belief-commitments consist not only of one's explicit beliefs but include their consequences as well, once logical regimentation has established the consistency requisite for "*rational* belief."

2. Ramsey 1931, p. 247.

3. The relevant considerations regarding epistemic logic are canvassed in Rescher, "Epistemic Logic," in Jacquette 2002, pp. 478–490.

4. That stipulations can predominate even over these will become clear in the discussion of *per impossibile* reasoning in section 6 of chapter 12.

5. Plausibility is a matter of classification and not of mensuration. Thus the discrepancy in *modus operandi* between probability and plausibility means that in view of the Lottery Paradox some statements of fairly high probability ("Some statement I believe is false") must nevertheless be accorded zero plausibility. On the nature and modus operandi of plausibility see Rescher 1976. The history of the weakest link principle goes back to the thesis of the Greek philosopher Speusippus that the modality of the conclusion of a modal syllogism is the same as that of its weakest premise.

Chapter 6 Inferentially Insuperable Boundaries and Homogeneous Conditionals (Pages 67–72)

1. The condition that not $\vdash r$ enters in to assure that the premises play a role in establishing r.

2. To be sure, modally iterative propositions must be excluded here. "It is actually the case that p is merely possible" otherwise affords an actualistic premise that yields the merely possibilistic conclusion "p is merely possible." Moreover, the situation changes when we resort to "propositional attitudes": the reality-descriptive and clearly factual statement "John regrets that Peter didn't come although he could have" entails the thesis "Peter didn't come although he could have," which affirms an unrealized possibility. Contrariwise, it is also possible to infer facts from statements of mere possibility: from the claim "It is possible (but not actual) that a certain Dr. Frankenstein brought Nostradamus back to life in 1810" we can infer the factual truth that Nostradamus had died prior to 1810.

3. John F. Whippel, "The Relationship between Essence and Existence in Late-Thirteenth-Century Thought," in Morewedge 1982, pp. 136–138.

4. After J. S. Mill, who in his *Utilitarianism* in effect maintained that what is preferable (i.e., deserving of preference) can be deduced from what people *de facto* prefer.

5. To be sure, in the order of *causal* explanation there need be no kinship between cause and effect—one sort of thing can in theory cause something of a very different sort. After all, the whole object of the idea of efficient causality is to possibilize connections where conceptual linkages do not exist. But in the order of conceptual explication like has to be treated in terms of like. The *occurrences* of overall events can certainly be explained causally with reference to physical occurrences. There is nevertheless no way of inferring the conceptual nature of the mental in terms of reference drawn wholly from the realm of the physical. At this *hermeneutic* level there is an inferential barrier that is as unbridgeable as any.

Chapter 7 Counterfactual Conditionals and Their Problematic Nature (Pages 73–88)

1. The pioneer studies are Chisholm 1946 and Goodman 1946. See also Schneider 1953.

2. Sometimes what looks like a counterfactual conditional is only so in appearance. Thus consider "If Napoleon and Alexander the Great were fused into a single individual, what a great general that would be!" What is at issue here is not really a counterfactual based on the weird hypothesis of a fusion of two people into one. Rather, what we have is merely a rhetorically striking reformulation of the truism that "Anybody with all of the military talents of Napoleon and of Alexander combined is certainly a great general."

3. "Supposition on the part of the Creator would be ridiculous, for supposition implies doubt." Quoted in Guillaume 1934, p. 62. As noted above, conditionals whose antecedents are seen as true—factual conditionals, that is—are generally formulated with *since* rather than *if*.

4. Herodotus, *History*, 2.20: 2–3. For more information on the historical background of counterfactuals see Rescher, "Thought Experimentation in Precocratic Philosophy," in Horowitz and Massey 1991, pp. 31–41.

5. Translated in part in Kretzman and Stump, 1988, pp. 389–412.

6. On Burley's principle see also Spade 1982, 1992.

7. Chisholm 1946. Ramsey's original publications of the 1920s are reprinted in Ramsey 1978a.

8. The treatment of suppositions presented in this chapter was initially set out in Rescher 1961 and subsequently developed in Rescher 1964.

9. See Hansson 1995, pp. 15–17.

10. On this issue see Rescher 1964 and Levi 1996. The sort of logical or quasi-logical approach envisioned by various 1940s theorists such as Chisholm 1946 and Goodman 1946 is foredoomed to failure as the subsequent unfolding of discussion during the second half of the twentieth century has made only too clear.

11. This point was stressed in Rescher 1964. It has also been argued in Hansson 1995.

Chapter 8   Salience and Questioner's Prerogative (Pages 89–102)

1. See Quine 1974, p. 21.

2. This issue of questioner's prerogative is yet another point of contact between the theory of counterfactuals and the medieval theory (and practice) of disputation. See Spade 1982, 1992.

3. See Quine 1960.

4. The Oswald-Kennedy example is due to Lewis 1973a, p. 3, who employs it to very different ends. Note that if both of those bracketed beliefs are seen as having equal salience in the question-context at hand, then we can do no better than the disjunctive conditional:

If Oswald had not assassinated Kennedy then either someone else did it or he would not have been assassinated at all.

5. Note that as more and more accepted truths are taken to be salient this range of choice increases so that the consequent of the conditional becomes more indefinite.

6. On the difference between truth conditions and conditions of appropriate use or assertibility see chapter 3 of Rescher 1998. The contrast at issue goes back at least to Adams 1965, pp. 171–72. Adams however, inclines to construe assertability as a matter of probability. This is deeply problematic in the light of nonconjectivity and the Lottery Paradox.

Chapter 9   On Validating Counterfactuals (Pages 103–138)

1. Sometimes what looks like a counterfactual conditional is only so in appearance. Thus consider "If Napoleon and Alexander the Great were fused into a single individual, what a great general that would be!" What is at issue here is not really a counterfactual based on the weird hypothesis of a fusion of two people into one. Rather, what we have is merely a rhetorically striking reformulation of the truism that "Anybody with all of the military talents of Napoleon and of Alexander combined, is certainly a great general."

2. Chisholm 1946. Ramsey's original publications of the 1920s are reprinted in Ramsey 1978.

3. For example, in discussing thought experiments intended to demonstrate absolute motion, the physicist Ernst Mach maintained that: "When experimenting in thought, it is permissible to modify *unimportant* facts in order to bring out new features in a given case [but not important ones]" (Mach 1960, p. 341). (I owe this reference to Paul Ehrlich.) For further historical data see the postscript to this chapter.

4. Not only does this have the backing of a sound rationale of theoretical analysis, but a whole host of empirical studies coalesce to indicate that people in fact think in this way. See Revlis, Lipkin, and Hayes 1971; Revlis and Hayes 1972; Kelley and Michaela 1980; Kahneman and Twersky 1982; Johnson 1986; Kahneman and Miller 1986; Roese and Olsen 1993; Gavanski and Wells 1989; Miller, Turnbull, and McFarland 1990; as well as the many relevant studies cited by these authors.

5. On this issue see N. Rescher, *Plausible Reasoning* (Van Gorcum: Assen, Netherlands, 1976).

6. This issue is deliberated in Adams 1970.

7. On this issue compare Rescher 1961, 1964; Adams 1970; and Lewis 1976.

8. See Goodman 1947.

9. See Lewis 1973, p. 1.

10. Wilson 1986, pp. 260–273.

11. In Rescher 1964, *Hypothetical Reasoning*.

12. Ibid., p. 35.

13. To be sure, as Pierre Duhem insisted a theory-observation clash will in general involve a plausibility of participating theories, so that it will not be clear which particular theory will have to be jettisoned. See Duhem 1890.

14. Reid 1785, VI, iv, p. 570.

15. Reid 1764, I, v (Hamilton edition, p. 102b; Brooks edition, p. 21).

16. See Fine 1984, esp. p. 85. The maxim is articulated in line with David Hilbert's endeavor to demonstrate the consistency of set theory on a more stringent non-set-theoretical basis.

17. See Rescher 1961, 1973.

18. For empirical confirmation of generality prioritization in counterfactual reasoning as people actually conduct it see Revlis, Lipkin, and Hayes 1971 as well as Revlis and Hayes 1972.

19. See Bunge 1979, p. 88.

20. See Unger 1982.

21. See Kripke 1980, esp. p. 42.

22. The context-dependent nature of the project of conflict resolution means that the aporetic approach is able to unify important aspects of the theory of reasoning in very different domains (proof theory, empirical inquiry, hypothetical reasoning, philosophical reasoning) within a single overarching integrating perspective. This unification patently integrates my approach to these various issues in such books as *Hypothetical Reasoning* (1967), *Plausible Reasoning* (1974), *Empirical Inquiry* (1982), and *The Strife of Systems* (1985), and thereby unifies in one single synoptic perspective the pragmatic tendency of my overall position.

23. Rescher 1961.

24. Revlis and Hayes 1972; Revlis, Lipkin, and Hayes 1971. See also Braine and O'Brian 1991 and Roese and Olson 1995, p. 4.

25. See Lewis 1973, 1977; Sorensen 1992; and Unger 1982. Unger sees the matter as one of "the psychology of thought experimentation" and insight into a psychological "residence of abandonment."

26. On generality-precedence see also the discussion of plausibility prioritization in Rescher 2001, pp. 47–51.

Chapter 10 Further Complications of Counterfactuality (Pages 139–148)

1. Any "closest possible worlds" analysis of counterfactuals will run into problems in these cases where there just are not such worlds.

Chapter 11   Some Logical Features of Counterfactuals (Pages 149–160)

1. On nonmonotonic inference see "Common-Sense Reasoning" in *The Routledge Encyclopedia of Philosophy* (London: Routledge, 2000) as well as Rescher 1980 and also Kyburg and Teng 2001.

2. On these issues see Harper 1981, pp. 4–7.

3. See Woods 1997, p. 124.

4. See Stalnaker 1968 and Lewis 1973a.

5. Compare the discussion of Geach's thesis in section 1 of chapter 1 above. On relevant issues see also Appiah 1981, p. 205–210; Gibbard 1981, pp. 234–238; Edgington 1995, pp. 180–184; Jackson 1977, pp. 127–137; and Woods 1997, pp. 58–68, 120–124.

6. The only way we could now make sense of nested $\vdash$ is by way of subordinate deductions proceeding from assumptions within assumptions. And although $(p \vdash q) \vdash r$ may seem problematic, various of its instantiations such as $(p \vdash q) \vdash (\sim q \vdash \sim p)$ will be unproblematically tenable.

Chapter 12   Variant Analyses of Counterfactuals (Pages 161–176)

1. Ramsey 1978b. See Sahlin 1990, p. 121.

2. See the reference given in Crocco, del Cerro, and Herzig 1995, p. 148.

3. Ibid.

4. See Furhman and Levi 1969. However, these authors, to mitigate the difficulties they envision, introduce so complex a variant of the "Inductive Ramsey Test" that the effect is one of employing a steamroller to crack a nut. A theoretical analysis of counterfactuals must not lose sight of the fact that they are a commonplace resouce of ordinary discourse.

5. See Pollock 1976.

6. Stalnaker 1968, pp. 33–34 (Jackson reprint).

7. For Lewis, when $\sim p$ is a (contingent) truth then p counterfactually implies q iff some world at which $p \& q$ holds is closer to the actual world than any world in which $p \& \sim q$ holds.

8. The theoretical basis of possible-world avoidance is set out in considerable detail in Rescher 2003.

9. Ibid.

10. Compare Chihara 1998, p. 79.

11. See Harper 1981, p. 9.

12. On this issue compare also Fine 1975.

13. See Lewis 1979 (1986 reprint), pp. 43–48. Lewis speaks of these fictions as a matters of "weights or priorities." But these are very different matters, since weighing counterfactuals blending while prioritization looks to a lexical ordering. Lewis's account has other internal difficulties as well. (See Elga 2002.)

14. It should, however, be noted that the indicated result hinges on regarding these regularities as lawful. In the absence of this basis for prioritization, the counterfactual supposition involves a plurality of undifferentiated alternatives, where "there is just no saying" where that x would be.

15. This is an oversimplification. On Lewis's approach we would need to construct a possible world to implement the scenario at issue—a task which could in theory be achieved in many different ways and in practice in none.

16. See Rescher 1961, 1964.

17. On these issues, see Rescher 2003.

18. See Adams 1966, 1975; Sanford 1989, and the critique of this approach present in Lewis 1976.

19. See chapter 4 above.

Chapter 13   Historical Counterfactuals (Pages 177–184)

1. When the British Prime Minister Lord Rosebery was installed as Lord Rector of Glasgow University in the late 1800s, he gave a widely reported inaugural address in which he stated that if the American Revolution had been averted there would have resulted "a self-adjusting system of representation and in due time, when the majority of seats in the Imperial Parliament should belong to the section beyond the seas, the seat of empire would have been moved solemnly across the Atlantic" (Rosebery 1899, p. 286).

2. Problematic or not, this sort of thing has—rather surprisingly—become increasingly popular among historians in recent days. See Cowley 1998 and Ferguson 1999. Greenhill Books of London has launched an entire series along these lines: See Macksey 1998, 1999; Sobel 1997; P. G. Tsouras 2000; and P. G. Tsouras 1997. See also Thomsen and Greenberg 2002.

3. See Lewis 1979, Bennett 1984, and Adams 1993.

4. There have, to be sure, been cumbersome effort to handle counterfactuals by Lewisian possible-world semantics.

Chapter 14   *Per impossibile* Counterfactuals and *Reductio ad absurdum* Conditionals (Pages 185–194)

1. On this argumentation and its historic background see Heath 1921.

2. See Thompson 1954.

3. A somewhat more interesting mathematical example is as follows: If, *per impossibile*, there were a counterexample to Fermat's last theorem, there would be infinitely many counterexamples, because if $xk + yk = zk$, then $(nx)k + (ny)k = (nz)k$, for any k.

4. Here the necessity at issue can of course be either logico-conceptual or physical.

5. On these issues, see Rescher 1994.

6. Quoted James 1890, p. 80.

Chapter 15   Problems with Possible Worlds (Pages 195–216)

1. "A possible world, then, is a possible state of affairs—one that is possible in the broadly logical sense" (Plantinga 1974, p. 44).

2. Some logicians approach possible worlds by way of possible-world characterizations construed as collections of statements rather than objects. And there is much to be said on behalf of such an approach. But it faces two big obstacles: (1) not every collection of (compatible) statements can plausibly be said to constitute a world; rather, (2) only those can do so which satisfy an appropriate manifold of special conditions intending that any "world-characterizing" set of propositions must both inferentially closed and descriptively complete by way of assuring that any possible contention about an object is either true or false.

3. Van Inwagen 1989, pp. 419ff., questions whether we can uniquely ostend "the" world we live in, since he holds that actual individuals can also exist in other possible worlds. But this turns matters upside down. For unless one has a very strange sort of finger, its here-and-now pointing gesture does not get at things in those other worlds. There is no way of getting lost en route to a destination to which we cannot go at all.

4. Compare Felt 1983.

5. Leibniz, to be sure, was entitled to conjure with alternative possible worlds because they were, for him, theoretical resources as instances of God's *entia rationis*. Were one to ask him where possible worlds are to come from, he would answer "Only God knows." As that is exactly correct—only God does know. We feeble humans have no way to get there from here. (On Leibniz's theory of possibility see Mates 1986.)

6. Lewis 1986a, p. 2. The many-worlds theory of quantum mechanics projected by Everett and Wheeler can also be considered in this connection. Other "modal realists" (as they are nowadays called) include not only Leibniz but Robert Adams (see his 1974), and Robert Stalnaker (see his 1984).

7. Despite abjuring a spatial metaphor, Lewis's theory in one of its versions required a metric to measure how near or far one possible world is from another. This leads to hopeless problems. Is a world with two-headed cats closer to or more remote from ours than one with two-headed dogs?

8. Lewis 1973a, pp. 85–86.

9. On this Lewis–Lycan controversy—see Lycan 2001, pp. 85–86—the present deliberations come down emphatically on Lycan's side.

10. Lewis 1986a devotes to this problem a long section (pp. 108–115) entitled "How Can We Know?" It is the most unsatisfactory part of his book, seeing that what it offers is deeply problematic, owing to its systematic slide from matters of knowledge regarding possibility *de dicto* to existential commitments *de re*.

11. Mackie 1973b, p. 84.

12. Powers 1976, p. 95.

13. Statements of the format $(\exists w)Fw$ are systematically false unless Fw^*. But this of course does not hold for $T(\exists w)Fw$.

14. On this issue see Chisholm 1967a, Lewis 1968, Plantinga 1974 (chap. 6), Forbes 1985, and Chihara 1998 (chap. 2). The problem, of course, vanishes once we turn from possible worlds to the schematic scenarios as assumption and supposition. Here objects in different contexts are the same just exactly when this is stipulated in the formative hypotheses of the case.

15. Compare the discussion of section 2 of chapter 12 above.

16. See Chisholm 1967a.

17. Chihara 1998, p. 90.

18. Lycan 1979, pp. 295–296.

19. Stalnaker 1968, pp. 111–112.

20. In the semantic literature, such "worlds" are also called "partial" or "incomplete," but this of course concedes that they are not really *worlds* at all.

21. Stalnaker 1968, pp. 111–112.

22. See Plantinga 1974.

23. See, for example, Zalta 1988.

24. Compare Quine 1948.

25. Recombinatory suppositions of this sort are the guiding idea behind David Armstrong's approach to nonexistent possible worlds as nonactual recombinations of actual objects in Armstrong 1989.

26. See Lycan 1979.

27. See Lewis 1973a.

28. Latter-day combinationist theories along these lines are offered by Cresswell 1972, Skyrms 1981, and Armstrong 1989. Regarding such possibilia arising from redistribution or recombination, see Lewis 1986a, and also Rosenkranz 1980, 1984, 1985–1986, and esp. 1993.

29. And what about *counting* possible worlds? Counting anything, be it worlds or beans, presupposes identifying the items to be counted. What we cannot individuate we cannot count either. Given that one cannot tell just where one cloud leaves off and where another begins, one cannot count the clouds in the sky. Given that one cannot tell one idea from the rest one cannot count how many ideas a person has in an hour. And the same story goes for possible worlds. If we cannot identify possible worlds, we cannot possibly count them. How many of them are there? God only knows. As far as we are concerned, possible worlds are literally uncountable. And this is so not because they are too numerous but because they lack the critical factor of individuation/identification. Accordingly, we have little alternative but to see the question of quantification as inappropriate—effectively meaningless—seeing that it rests on presupposing our doing something that is in principle impossible for us.

30. See Mondadori and Martin 1976 (Loux 1979), pp. 245–246.

31. On viable conditionals with impossible antecedents, see Lewis 1973a, pp. 24–26. After briefly toying with the idea of impossible possible worlds, Lewis proposes to regard such counterfactuals as uniformly "vacuously true," an approach which would saddle us with all sorts of bizarre counterfactuals that no one would want to assert.

32. Lycan 1984, p. 17.

References

Adams, Ernest W. 1965. "The Logic of Conditionals." *Inquiry* 8: 166–197.

———. 1966. "Probability and the Logic of Conditionals." In *Aspects of Inductive Logic*, ed. J. Hintikka and P. Suppes, pp. 256–316. Amsterdam: North-Holland.

———. 1970. "Subjective and Indicative Conditionals." *Foundations of Language* 6: 39–94.

———. 1975. *The Logic of Conditionals: An Application of Probability to Deductive Logic.* Dordrecht: D. Reidel.

———. 1976. "Prior Probabilities and Counterfactual Conditionals." In *Foundations of Probability Theory, Statistical Inference and Statistical Theories of Science*, ed. W. L. Harper and C. A. Hooker. Dordrecht: Reidel.

———. 1977. A Note Comparing Probabilistic and Modal Logics of Conditionals. *Theoria* 41.

———. 1979. "Primitive Thisness and Primitive Identity." *Journal of Philosophy* 76: 5–26.

———. 1981. "Truth, Proof, and Conditionals." *Pacific Philosophical Quarterly* 62: 323–339.

———. 1987. "On Meaning of the Conditional." *Philosophical Topics* 15: 5–22.

———. 1993. "On the Rightness of Certain Counterfactuals." *Pacific Philosophical Quarterly* 74: 1–10.

———. 1998. *A Primer of Probability Logic.* Stanford: CLSI Publications.

Adams, E. W., and H. P. Levine. 1975. "On the Uncertainties of Transmitted from Premises to Conclusions in Deductive Inferences." *Synthese* 30.

Adams, Robert M. 1974. "Theories of Actuality." *Noûs* 81: 211–233.

Akatsuka, Noriko. 1986. "Conditionals Are Discourse-Bound." In *On Conditionals*, Traugott et al., pp. 333–352, Cambridge: Cambridge University Press.

Anderson, Alan Ross. 1951–52. "A Note on Subjunctive and Counterfactual Conditionals." *Analysis* 12: 35–38.

Anderson, Alan Ross, and Nuel Belnap. 1975. *Entailment: The Logic of Relevance and Necessity.* Princeton, N.J.: Princeton University Press.

Anderson, John. 1952. "Hypotheticals." *Australasian Journal of Philosophy* 30.

Anscombe, G. E. M. 1969. "Causality and Extensionality." *Journal of Philosophy* 66.

Appiah, Kwame. 1981. "Conditions for Conditionals." Unpublished Ph.D. thesis, Cambridge University, Cambridge.

———. 1982. "Conversation and Conditionals." *Philosophical Quarterly* 32.

———. 1985. *Assertion and Conditionals.* Cambridge: Cambridge University Press.

Aqvist, Lennart. 1971. *Modal Logic with Subjective Conditionals and Dispositional Predicates.* Uppsala: Uppsala Universitet.

Arló-Costa, Horacio. 2001. "Baysian Epistemology and Epistemic Conditionals." *Journal of Philosophy* 98: 555–593.

Armstrong, David M. 1989. *A Combinatorial Theory of Possibility.* Cambridge: Cambridge University Press.

Athanasiadou, Angeliki, and René Dirven (eds.). 1997. *On Conditionals Again.* Amsterdam: John Benjamins.

Austin, J. L. 1961. "Ifs and Cans." In *Philosophical Papers*, ed. J. O. Urmson and G. J. Warnock. Oxford: Clarendon Press.

Ayers, M. R. 1965. "Counterfactuals and Subjunctive Conditionals." *Mind* 74: 347–364.

Baker, A. J. 1967. "If and ⊃." *Mind* 76.

Barker, John A. 1969. *A Formal Analysis of Conditionals.* Carbondale, Ill.: Southern Illinois University Press.

Bennett, Jonathan. 1974. "Counterfactuals and Possible Worlds." *Canadian Journal of Philosophy* 4: 381–402.

———. 1976. *Linguistic Behaviour.* Cambridge: Cambridge University Press.

———. 1982. "Even If." *Linguistics and Philosophy* 5: 403–408.

———. 1984. "Counterfactuals and Temporal Direction." *Philosophical Review* 93: 57–91.

———. 1988. "Farewell to the Phlogiston Theory of Conditionals." *Mind* 97: 509–527.

———. 1995. "Classifying Conditionals: The Traditional Way Is Right." *Mind* 104: 331–344.

Bigelow, J. C. 1976. "If-Then Meets the Possible Worlds." *Philosophia* 6.

Bowie, G. L. 1979. "The Similarity Approach to Counterfactuals: Some Problems." *Noûs* 13: 477–498.

Bradley, F. H. 1922. *The Principle of Logic*, 2nd ed. Oxford: Clarendon Press.

Braine, M. D., and D. P. O'Brian. 1991. "A Theory of If: A Lexical Entry, Reasoning Program and Pragmatic Principles."*Psychological Review* 98: 182–203.

Brown, Robert, and John Watling. 1950. "Counterfactual Conditionals." *Mind* 57: 222–233.

Bunge, M. 1979. *Epistemology.* Dordrecht: Reidel.

Carlstrom, I. F., and C. S. Hill. 1978. "Rev: Adams, E. W." *The Logic of Conditionals: Philosophy of Science*, vol. 45.

Chandler, H. S. 1978. "What Is Wrong with Addition of an Alternate?" *Philosophical Quarterly* 28.

Chellas, B. F. 1975. "Basic Conditional Logic." *Journal of Philosophical Logic* 4: 133–153.

Chihara, Charles S. 1998. *The Worlds of Possibility: Modal Realism and the Semantics of Modal Logic.* Oxford: Clarendon Press.

Chisholm, R. M. 1946. "The Contrary-to-Fact Conditional." *Mind* 55: 289–307.

———. 1955. "Law Statements and Counterfactual Inference." *Analysis* 15: 97–105. Reprinted in Sosa 1975, pp. 167–155.

———. 1967a. "Identity through Possible Worlds: Some Questions." *Noûs* 1: 1–8. Reprinted in Loux 1979, pp. 80–87.

———. 1967a. "Meinong, Alexius." In *The Encyclopedia of Philosophy*, ed. P. Edwards, vol. 5, pp. 261–263. New York: Macmillan.

Clark, M. 1971. "Ifs and Hooks." *Analysis* 32.

———. 1974. "Ifs and Hooks: A Rejoinder." *Analysis* 34.

Cohen, L. J. 1971. "Some Remarks on Grice's Views about the Logical Particles of Natural Language." In *Pragmatics of Natural language*, ed. J. BarHillel. Dordrecht: Reidel.

———. 1977. "Can the Conversationalist Hypothesis Be Defended?" *Philosophical Studies* 31.

Cooper, W. S. 1968. "The Propositional Logic of Ordinary Discourse." *Inquiry* 11.

———. 1978. *The Foundations of Logico-Linguistics*. Dordrecht: Reidel.

Cowley, Robert (ed.). 1998. *What If*. New York: G. P. Putnam's Sons.

Cresswell, M. J. 1972. "The World Is Everything That Is the Case." *Australian Journal of Philosophy* 50: 1–13. Reprinted in Loux 1979, pp. 129–145.

Crocco, G., Luis Fariñas del Cerro, and A. Herzig (eds.). 1995. *Conditionals: From Philosophy to Computer Science*. Oxford: Oxford University Press.

Dale, A. J. 1972. "The Transitivity of 'If, Then.' " *Logique et Analyse* 15.

———. 1974. "A Defense of Material Implication." *Analysis* 34.

Dancygier, Barbara. 1998. *Conditionals and Prediction*. Cambridge: Cambridge University Press.

Davidson, D., and G. Harman. 1972. *Semantics of Natural Language*. Dordrecht, Reidel.

Davis, Wayne A. 1979. "Indicative and Subjunctive Conditionals." *Philosophical Review* 88: 544–564.

———. 1980. "Jackson on Counterfactuals." *Australasian Journal of Philosophy* 58: 62–65.

Downing, P. B. 1958–1959. "Subjunctive Conditionals, Time Order, and Causation." *Proceedings of the Aristotelian Society* 59: 125–140.

Dudman, V. H. 1983. "Tense and Time in English Verb Clusters of the Primary Pattern." *Australian Journal of Linguistics* 3: 35–44.

———. 1984. "Conditional Interpretations of 'If'-Sentences." *Australian Journal of Linguistics* 4: 143–204.

———. 1984. "Parsing 'If'-Sentences." *Analysis* 44: 145–153.

———. 1987. "Appiah on 'If.' " *Analysis* 47: 74–79.

———. 1988. "Indicative and Subjunctive." *Analysis* 48: 113–122.

Duhem, Pierre. 1890. *La théorie physique, son objet et sa structure*. Paris: A. Hermann.

Dummett, Michael. 1973. *Frege*. (2nd ed., 1981.) Cambridge, Mass.: Harvard University Press.

———. 1981. *Philosophy of Language*, 2nd ed. Cambridge, Mass.: Harvard University Press.

Edgington, Dorothy. 1986. "Do Conditionals Have Truth-Conditions?" *Critica* 18, no. 52: 3–30. Reprinted in Frank Jackson (ed.), *Conditionals* (q.v.), pp. 176–201.

———. 1995. "On Conditionals." *Mind* 104: 329–335.

———. 1991. "The Mystery of the Missing Matter of Fact." *Proceedings of the Aristotelian Society (suppl. vol.)* 65: 185–209.

———. "Conditionals." *Stanford Encyclopedia of Philosophy*. Http://plato.stanford.edu.

Edwards, S. 1974. "A Confusion about 'If . . . Then.' " *Analysis* 34.

Eells, Ellery, and Brian Skyrms (eds.). 1994. *Probability and Conditionals: Belief Revision and Rational Decision*. Cambridge: Cambridge University Press.

Elga, Adam. 2002. "Statistical Mechanics and the Asymmetry of Counterfactual Dependence." *Philosophy of Science (suppl. vol.)* PSA 2000.

Ellis, Brian. 1969. "An Epistemological Concept of Truth." In *Contemporary Philosophy in Australia*, ed. R. Brown and C. D. Rollins. London: Allen & Unwin.

———. 1979. *Rational Belief Systems*. Totowa, N.J.: Rowman & Littlefield.

————. 1984. "Two Theories of Indicative Conditionals." *Australasian Journal of Philosophy* 62: 50–66.

Ellis, B., F. Jackson, and R. Pargetter. 1977. "An Objection to Possible World Semantics for Counterfactuals." *Journal of Philosophical Logic* 6: 355–357.

Epstein, Richard L. 2002. *Five Wages of Saying "Therefore."* Belmont, Calif.: Wadsworth.

Etchemendy, J. 1990. *The Concept of Logical Consequence.* Cambridge, Mass.: Harvard University Press.

Felt, James W. 1983. "Impossible Worlds." *International Philosophical Quarterly* 23: 251–265.

Ferguson, Niall (ed.). 1999. *Virtual History.* New York: Basic Books.

Fillenbaum, Samueal. 1986. "The Use of Conditionals in Inducements and Deterrents." In *On Conditionals*, ed. Traugott et al., pp. 179–196. Cambridge: Cambridge University Press.

Finch, H. A. 1957–1958. "An Explication of Counterfactuals by Probability Theory." *Philosophy and Phenomenological Research* 18.

————. 1959–1960. "Due Care in Explicating Counterfactuals: a Reply to Mr. Jeffery." *Philosophy and Phenomenological Research* 20.

Fine, A. 1984. "The National Ontological Attribute." In *Scientific Realism*, ed. Jarret Lephon, pp. 83–107. Berkeley and Los Angeles: University of California Press.

Fine, K. 1975. "Review of Lewis's *Counterfactuals.*" *Mind* 84: 451–458.

Forbes, Graeme. 1985. *The Metaphysics of Modality.* Oxford: Oxford University Press.

Ford, Cecilia E., and Sandra A. Thompson. 1986. "Conditionals in Discourse: A Text-Based Study from English." In Traugott et al. 1986, pp. 353–372.

Frege, Gottlob. 1966. *Begriffschrift.* In *Translations from the Philosophical Writing of Gottlob Frege*, ed. Peter Geach and Max Black. Oxford: Basil Blackwell.

Fuhrmann, André, and Isaac Levi. 1969. "Undercutting the Ramsey Test for Conditionals." In *The Logic of Strategy*, ed. Christina Biccieri et al. Oxford: Oxford University Press.

Gable, Lou. 2000. *The Blackwell Guide to Philosophical Logic.* Oxford: Blackwell.

Gärdenfors, P. 1978. "Conditionals and Changes of Belief." In *The Logic and Epistemology of Scientific Change*, vol. 20 (2–4) of *Acta Philosophica Fennica*, ed. I. Niiniluoto and R. Tuomela. Amsterdam: North Holland.

————. 1986. "Belief Revision and the Ramsey Test for Conditionals." *Philosophical Review* 95: 81–93.

————. 1988. *Knowledge in Flux.* Cambridge, Mass.: MIT Press.

Gavanski, I., and G. L. Wells. 1989. "Counterfactual Processing of Normal and Exceptional Events." *Journal of Experimental Social Psychology* 25: 314–325.

Gazdar, G. 1979. *Pragmatics: Implicature, Presupposition, and Logical Form.* London: Academic Press.

Gibbard, Allan. 1981. "Two Recent Theories of Conditionals." In Harper, Pearce, and Stalnaker 1981, pp. 211–247.

Gibbard, Allan, and W. L. Harper. 1978. "Counterfactuals and Two Kinds of Expected Utility." In *Foundations and Applications of Decision Theory*, ed. C. A. Hooker, J. J. Leach, and E. F. McLennen. Dordrecht: Reidel.

Gibbins, P. F. 1979. "Material Implication, the Sufficiency Condition and Conditional Proof." *Analysis* 39.

————. 1984. "Why the Distributive Law Is Sometimes False." *Analysis* 44: 64–67.

Goodman, Nelson. 1946. "The Problem of Counterfactual Conditionals." *Journal of Philosophy* 44: 113–128. Reprinted in Goodman 1954, pp. 3–27, and in Jackson 1987, pp. 9–27.

————. 1954. *Fact, Fiction, and Forecast.* (4th ed., 1983.) Cambridge, Mass.: Harvard University Press.

Govier, Trudy. 1971. "What's Wrong with Slippery Slope Arguments?" *Canadian Journal of Philosophy* 8: 186–191.

Grandy, R. E. 1980. "Ramsey, Reliability, and Knowledge." In *Prospects for Pragmatism*, ed. D. H. Mellor. Cambridge: Cambridge University Press.

Grant, W. Mathews. 2000. "Counterfactuals of Freedom, Future Contingents, and the Grounding Objection to Middle Knowledge." *Proceedings of the American Catholic Philosophical Association* 74: 307–323.

Grice, H. P. 1975. "Logic and Conversation." In *The Logic of Grammar*, ed. D. Davidson and G. Harman. Encino, Calif.: Dickenson Publishing.

———. 1989. "Indicative Conditionals." In his *Studies in the Way of Words*. Cambridge, Mass.: Harvard University Press.

Guillaume, Alfred (ed.). 1934. *The "Summa Philosophiae" of Al-Sharastānī*. Oxford: Clarendon Press.

Haak, Susan. 1978. *Philosophy of Logics*. Cambridge: Cambridge University Press.

———. 1996. *Deviant Logic, Fuzzy Logic*. Chicago: University of Chicago Press.

Hacking, Jane F. 1998. *Coding the Hypothetical*. Philadelphia: John Benjamins.

Haiman, John. 1978. "Conditionals Are Topics." *Language* 54: 564–589.

Hampshire, Stuart. 1948. "Subjective Conditionals." *Analysis* 9: 9–14. Reprinted in *Philosophy and Analysis*, ed. M. Macdonald, pp. 204–210. Oxford: Clarendon Press, 1954.

Hansson, S. O. 1995. "The Emperor's New Cloths: Some Recurring Problems in the Formal Analysis of Counterfactuals." In *Conditionals: From Philosophy to Computer Science*, ed. G. Crocco, L. F. Fariñas del Cerro, and A. Herzig. Oxford: Clarendon Press.

Harper, William L. 1981. "A Sketch of Some Recent Developments in the Theory of Conditionals." In Harper, Pearce, and Stalnaker 1981, pp. 3–38.

———. 1976. "Ramsey Test Conditionals and Iterated Belief Change." In Harper and Hooker 1976, pp. 73–113.

Harper, W. L., and C. A. Hooker (eds.). 1976. *Foundations of Probability Theory, Statistical Inference, and Statistical Theories of Science*, vol. 1. Dordrecht: D. Reidel.

Harper, W. L., G. Pearce, and R. Stalnaker (eds.). 1981. *Ifs: Conditionals, Belief, Decision, Chance, and Time*. Dordrecht: D. Reidel.

Harrison, H. 1968. "Unfulfilled Conditionals and the Truth of Their Constituents." *Mind* 77: 372–382.

Hazen, A., and Slote, M. 1979. "'Even If.'" *Analysis* 39.

Heath, T. L. 1921. *A History of Greek Mathematics*. Oxford: Clarendon.

Hilbert, David, and W. Ackermann. 1950. *Principles of Mathematical Logic*. New York: Chelsen.

Hintikka, J., and P. Suppes (eds.). 1966. *Aspects of Inductive Logic*. Amsterdam: North Holland.

Hiz, Henry. 1951. "On the Inferential Sense of Contrary-to-Fact Conditionals." *Journal of Philosophy* 48: 586–587.

Horowitz, Tamara, and Gerald J. Massey (eds.). *Thought Experiments in Science and Philosophy*. Lanham, Maryland: Rowman and Littlefield.

Isaac, Levi. 1996. *For the Sake of the Argument*. Cambridge: Cambridge University Press.

Jackson, Frank. 1977. "A Causal Theory of Counterfactuals." *Australasian Journal of Philosophy* 55: 3–21.

———. 1979. "On Assertion and Indicative Conditionals." *Philosophical Review* 88: 565–589. Reprinted in Jackson 1991, pp. 111–135.

———. 1980–1981. "Conditionals and Possibilia." *Proceedings of the Aristotelian Society* 81: 125–137.

———. 1990. "Classifying Conditionals I." *Analysis* 50: 134–147. Reprinted in Jackson 1998.

——— (ed.). 1991. *Conditionals.* Oxford: Clarendon Press.

———. 1998. *Mind, Method, and Conditionals.* London: Routledge.

Jacquette, Dale (ed.). 2002. *A Companion to Philosophical Logic.* Oxford: Blackwell.

James, William. 1890. *The Will to Believe and Other Essays in Pragmatic Philosophy.* New York: Longman's Green.

Jeffrey, R. C. 1959–1960. "A Note on Finch's 'An Explacation of Counterfactuals by Probability Theory.'" *Philosophy and Phenomenological Research* 20.

———. 1964. "If." *Journal of Philosophy* 61.

———. 1991. "Matter of Fact Conditionals." *Proceedings of the Aristotelian Society (suppl. vol.)* 65: 161–183.

Johnson, J. T. 1986. "The Knowledge of What Might Have Been: Affective and Attributional Consequences of Near Outcomes." *Personality and Social Psychology Bulletin* 12: 51–62.

Kahneman, D., and D. T. Miller. 1986. "Norm Theory: Comparing Reality to Its Alternatives." *Psychological Review* 93: 136–153.

Kahneman, D., and A. Twersky. 1982. "The Simulation Heuristic." In *Judgment Under Uncertainty*, ed. D. Kahneman, P. Slovic, and A. Twersky, pp. 201–208. Cambridge: Cambridge University Press.

Kazez, J. 1995. "Can Counterfactuals Save Mental Causation?" *Australian Journal of Philosophy* 73: 71–89.

Kelley, H. H., and J. Michaela. 1980. "Attribution Theory and Research." *Annual Review of Psychology* 31: 457–501.

Kirwan, C. A. 1987. *Logics and Argument.* London: Brill Academic.

Kneale, William. 1950. "Natural Laws and Contrary-to-Fact Conditionals." *Analysis* 10: 121–125. Reprinted in M. Macdonald, *Philosophy and Analysis*, Oxford: Clarendon Press, 1954.

Kneale, William, and Martha Kneale. 1985. *The Development of Logic.* Oxford: Clarendon Press.

Kretzmann, N. N., and E. Stump (eds.). 1988. *The Cambridge Translation of Medieval Philosophical Texts*, vol. 1: *Logic and Philosophy of Language.* Cambridge: Cambridge University Press.

Kripke, Saul. 1980. *Naming and Necessity.* Cambridge, Mass.: Harvard University Press.

Kuhn, T. S. 1981. "A Function for Thought Experiments in Science." In *Scientific Revolutions*, ed. Ian Hacking, pp. 6–27. Oxford: Clarendon Press.

Kyburg, H., Jr. 1970. "Conjunctivitis." In *Induction, Acceptance, and Rational Belief*, ed. M. Swain. Dordrecht: Reidel.

———, and Chon Man Teng. 2001. *Uncertain Inference.* Cambridge: Cambridge University Press.

Lance, Mark. 1991. "Probabilistic Dependence among Conditionals." *Philosophical Review* 100: 269–276.

Lehmann, S. K. 1979. "A General Propositional Logic of Conditionals." *Notre Dame Journal of Formal Logic* 20.

Leibniz, G. W. 1879. *Philosophische Schriften*, vol. 2. Ed. C. J. Gerhardt. Berlin: Weidmannsche Buchhandlung.

Lephon, Jarret (ed.). 1984. *Scientific Realism.* Berkeley and Los Angeles: University of California Press.

Levi, Isaac. 1996. *For the Sake of Argument.* Cambridge: Cambridge University Press.

Levinson, Stephen C. 1983. *Pragmatics*. Cambridge: Cambridge University Press.

Lewis, David K. 1968. "Counterpart Theory and Quantified Modal Logic." *Journal of Philosophy* 65: 113–126. Reprinted in Loux 1979, pp. 210–228.

———. 1970. "Completeness and Decidability of Three Logics of Counterfactual Conditionals." *Theoria* 37.

———. 1973a. *Counterfactuals*. Oxford: Blackwell.

———. 1973b. "Counterfactuals and Comparative Possibility." *Journal of Philosophical Logic* 2: 918–946. Reprinted in his *Philosophical Papers*, vol. 2 (Oxford: Oxford University Press, 1986).

———. 1976. "Probabilities of Conditionals and Conditional Probabilities." *Philosophical Review* 85: 297–315, 581–589. Reprinted in his *Philosophical Papers*, vol. 2 (Oxford: Oxford University Press, 1986), pp. 133–152; in Harper, Pearce, and Stalnaker 1981; and in Jackson 1987, pp. 76–101.

———. 1977. "Possible-World Semantics for Counterfactual Logics: A Rejoinder." *Journal of Philosophical Logic* 6.

———. 1979. "Counterfactual Dependence and Time's Arrow." *Noûs* 13: 57–91. Reprinted in his *Philosophical Papers*, vol. 2, pp. 32–66 (Oxford: Oxford University Press, 1986).

———. 1986a. *On the Plurality of Worlds*. Oxford: Basil Blackwell.

———. 1986b. "'Postscripts to' Counterfactual Dependent and Time's Arrow." In his *Philosophical Papers*, vol. 2. Oxford: Oxford University Press.

Loar, B. 1980. "Ramsey's Theory of Belief and Truth." In *Prospects for Pragmatism*, ed. D. H. Mellor. Cambridge: Cambridge University Press.

Loewer, B. 1976. "Counterfactuals with Disjunctive Antecedents." *Journal of Philosophy* 73: 531–536.

———. 1979. "Cotenability and Counterfactual Logics." *Journal of Philosophical Logic* 8: 99–115.

Loux, Michael J. 1979. *The Possible and the Actual*. Ithaca: Cornell University Press.

Lowe, E. J. 1979. "Indicative and Counterfactual Conditionals." *Analysis* 39: 139–141.

Lycan, William G. 1979. "The Trouble with Possible Worlds." In Loux 1979, pp. 274–316.

———. 1984. *Logical Form in Natural Language*. Cambridge, Mass.: MIT Press.

———. 2001. *Real Conditionals*. Oxford: Oxford University Press.

Mach, Ernst. 1905. *Erkenntnis und Irrtum*. Leipzig: J. A. Barth.

———. 1960. *The Science of Mechanics*, 6th ed. Trans. T. J. McCormack. La Salle, Ill.: Open Court.

Mackie, J. L. 1965. "Causes and Conditions." *American Philosophical Quarterly* 2: 245–264. Reprinted in Sosa 1975, pp. 15–38.

———. 1973a. "Counterfactuals and Causal Laws." In *Analytical Philosophy, 1st Series*, ed. R. J. Butler. Oxford: Clarendon Press.

———. 1973b. *Truth, Probability, and Paradox*. Oxford: Clarendon Press.

Macksey, Kenneth (ed.). 1998. *The Hitler Options: Alternate Decisions of World War II*. London: Greenhill Books.

Macksey, Kenneth. 1999. *Invasions: The Alternate History of the German Invasion of England, July 1940*. London: Greenhill Books.

Marcus, R. B. 1953. "Strict Implication, Deducibility, and the Deduction Theorem." *Journal of Symbolic Logic* 18.

Mates, Benson. 1986. *The Philosophy of Leibniz: Metaphysics and Language*. New York: Oxford University Press.

McCawley, J. D. 1974. "If and Only If." *Linguistic Inquiry*.

McDermott, Michael. 1996. "On the Truth-Conditions of Certain 'If'-Sentences." *Philosophical Review* 105: 1–37.

McGee, Vann. 1985. "A Counterexample to Modus Ponens." *Journal of Philosophy* 82: 262–271.

———. 1989. "Conditional Probabilities and Compounds of Conditionals." *Philosophical Review* 98: 485–542.

McKay, Thomas, and Peter van Inwagen. 1977. "Counterfactuals with Disjunctive Antecedents." *Philosophical Studies* 31: 353–356.

McLaughlin, Robert N. 1990. *On the Logic of Ordinary Conditionals.* Albany, N.Y.: State University of New York Press.

McLelland, J. 1971. "Epistemic Logic and the Paradox of the Surprise Examination." *International Logic Review* 3: 69–85.

Mellor, D. H. 1993. "How to Believe a Conditional." *Journal of Philosophy* 90: 233–248.

Miller, D. T., W. Turnbull, and C. McFarland. 1990. "Counterfactual Thinking and Social Perception: Thinking about What Might Have Been." In *Advances in Experimental Social Psychology*, vol. 23, ed. M. P. Zanna, pp. 305–331. Orlando, Florida: Academic Press.

Milne, Peter. 1997. "Bruno de Finetti and the Logic of Conditional Events." *British Journal for the Philosophy of Science* 48: 195–232.

Mondadori, Fabrizio, and Adam Martin. 1976. "Modal Realism: The Poisoned Pawn." *Philosophical Review* 85: 3–20. Reprinted in Loux 1979, pp. 235–252.

Morewedge, Parvitz (ed.). 1982. *Philosophies of Existence: Ancient and Medieval.* New York: Fordham University Press.

Myhill, John. 1953. "On the Interpretation of the sign ' ⊃ '." *Journal of Symbolic Logic* 18.

Nelson, J. O. 1966. "Is Material Implication Inferentially Harmless?" *Mind* 75.

Nute, Donald. 1975. "Counterfactuals and the Similarity of Worlds." *Journal of Philosophy* 72: 773–778.

———. 1980a. "Conversational Scorekeeping and Conditionals." *Journal of Philosophical Logic* 9: 153–166.

———. 1980b. *Topics in Conditional Logic.* Philosophical Studies Series in Philosophy, vol. 20. Dordrecht: Kluwer.

———. 1984. "Conditional Logic." In *Handbook of Philosophical Logic*, vol. 2, ed. D. Gabbay and F. Guenthner. Dordrecht: D. Reidel.

Parry, W. T. 1957. "Reexamination of the Problem of Counterfactual Conditionals." *Journal of Philosophy* 54: 85–94.

Plantinga, Alvin. 1974. *The Nature of Necessity.* Oxford: Oxford University Press.

Pollock, John L. 1976. *Subjunctive Reasoning.* Dordrecht: D. Reidel.

———. 1981a. "A Refined Theory of Counterfactauls." *Journal of Philosophical Logic* 10: 239–266.

———. 1981b. "Indicative Conditionals and Conditional Probability." In Harper, Pearce, and Stalnaker 1981.

Popper, K. 1949. "A Note on Natural Laws and So-Called 'Contrary-to-fact Conditionals.'" *Mind* 58.

———. 1959. "On Subjunctive Conditionals with Impossible Antecedents." *Mind* 68.

Powers, Larry. 1976. "Comments on Stalnaker's 'Propositions.'" In *Issues in the Philosophy of Language*, ed. A. F. MacKay and D. D. Merrill. New Haven: Yale University Press.

Putnam, Hilary. 1957. "Parry on Counterfactuals." *Journal of Philosophy* 54: 441–445.

———. 1983. "Foreword" to Goodman.

Quine, W. V. O. 1948. "On What There Is." *Review of Metaphysics* 2: 21–38. Reprinted in his *From a Logical Point of View*, 2nd ed. (New York: Harper Torchbooks, 1961), pp. 1–19; and in *Semantics and the Philosophy of Language*, ed. L. Linsky, pp. 189–206 (Urbana, 1952).

———. 1960. *Word and Object*. Cambridge, Mass.: MIT Press.

———. 1974. *Methods of Logic*, 3rd ed. London: Routledge and Kegan Paul.

———. 1975. "Mind and Verbal Dispositions." In *Mind and Language*, ed. S. Guttenplan. Oxford: Clarendon Press.

Ramsey, F. P. 1931. *The Foundations of Mathematics and Other Logical Essays*. Ed. R. B. Braithwaite. London: Methuen.

———. 1978a. *Foundations: Essays in Philosophy, Logic, Mathematics and Economics*, ed. D. H. Mellor. London: Routledge and Kegan Paul.

———. 1978b. "Law and Causality." In Ramsey 1978a.

Reade, Stephen. 1988. *Relevant Logic*. Oxford: Blackwell.

———. 1995. "Conditionals and the Ramsey Test." *Proceedings of the Aristotelian Society (suppl. vol.)* 69: 47–65.

Reid, Thomas. 1764. *An Inquiry into the Human Mind*. Also ed. by D. R. Brooks (University Park, Penn.: Pennsylvania State University Press), 1997.

———. 1785. *Essays on the Intellectual Powers of Man*. Edinburgh: John Bell.

Rescher, Nicholas. 1961. "Belief-Contravening Suppositions." *Philosophical Review* 87: 176–196. Reprinted in Sosa 1975, pp. 156–164.

———. 1962. "Quasi-Truth-Functional Systems of Propositional Logic." *Journal of Symbolic Logic* 27: 1–10.

———. 1964. *Hypothetical Reasoning*. Amsterdam: North Holland.

———. 1968a. "A Factual Analysis of Counterfactual Conditionals." *American Philosophical Quarterly Monograph* 2. Oxford: Basil Blackwell.

——— (ed.). 1968b. *Studies in Logical Theory*. Oxford: Blackwell.

———. 1973. *Conceptual Idealism*. Oxford: Basil Blackwell.

———. 1976. *Plausible Reasoning*. Assen: Van Gorcum.

———. 1980. *Induction*. Oxford: Basil Blackwell.

———. 1994. *Philosophical Standardism*. Pittsburgh: University of Pittsburgh Press.

———. 1998. *Communicative Pragmatism*. Lanham, Maryland: Rowan and Littlefield.

———. 2001. *Paradoxes: Their Roots, Range, and Resolution*. La Salle, Ill.: Open Court.

———. 2003. *Imagining Irreality: A Study of Unreal Possibilities*. Chicago and La Salle: Open Court.

Rescher, Nicholas, and Herbert Simon. 1966. "Cause and Counterfactual." *Philosophical Studies* 34: 323–340.

Revlis R., and J. R. Hayes. 1972. "The Primacy of Generalities in Hypothetical Reasoning." *Cognitive Psychology* 3: 268–290.

Revlis, R., S. G. Lipkin, and J. R. Hayes. 1971. "The Importance of Universal Quantifiers in a Hypothetical Reasoning Task." *Journal of Verbal Learning and Verbal Behavior* 10: 86–91.

Roese, N. J., and J. M. Olsen. 1993. "Self-Esteem and Counterfactual Thinking." *Journal of Personality and Social Psychology* 65: 199–206.

Roese, N. J., and J. M. Olson. 1995. *What Might Have Been: The Social Psychology of Counterfactual Thinking*. Mahweh, N.J.: Laurence Erlbaum.

Rosebery, A. P. P. (5th Earl). 1899. *Appreciations and Addresses*. London: Edwards.

Rosenkranz, Gary. 1980. "Reference, Intensionality, and Nonexistent Entities." *Philosophical Studien* 50.

———. 1984. "Nonexistent Possibles and Their Individuation." *Grazer Philosophische Studien* 22.

———. 1985–1986. "On Objects Totally Out of This World." *Grazer Philosophische Studies* 25/26.

———. 1993. *Hacceity: An Ontological Essay*. Dordrecht: Kluwer.

Ryle, Gilbert. 1950. "'If,' 'So,' and 'Because.'" In *Philosophical Analysis*, ed. Max Black. Ithaca, N.Y.: Cornell University Press.

Sahlin, Nils-Eric. 1990. *The Philosophy of F. P. Ramsey*. Cambridge: Cambridge University Press.

Sanford, David H. 1989. *If P, Then Q: Conditionals and the Foundations of Reasoning*. London: Routledge.

Schaur, Frederick. 1985. "Slippery Slopes." *Harvard Law Review* 99: 361–383.

Schlossberger, E. 1978. "Similarity and Counterfactuals." *Analysis* 38: 80–82.

Schneider, Erna F. 1953. "Recent Discussion of Subjective Conditionals." *Review of Metaphysics* 6: 623–649.

Sellars, W. S. 1956. "Counterfactuals, Dispositions, and the Causal Modalities." *Minnesota Studies in the Philosophy of Science*, vol. 2, ed. H. Feige, M. Scriven, and G. Maxwell, pp. 227–248. Minneapolis: University of Minnesota Press. Reprinted in Sosa 1975, pp. 126–146.

Sextus Empiricus. *Outlines of Pyrrhonism*.

Simon, Herbert A., and Nicholas Rescher. 1966. "Cause and Counterfactuals." *Philosophy of Science* 33: 323–340.

Skorupski, J. 1980. "Ramsey on Belief." In *Prospects for Pragmatism*, ed. D. H. Mellor. Cambridge: Cambridge University Press.

Skyrms, Bryan. 1981. "The Prior Propensity Account of Subjunctive Conditionals." In Harper, Pearce, and Stalnaker 1981, pp. 259–265.

———. 1980a. *Causal Necessity*. New Haven: Yale University Press.

———. 1980b. "Higher-Order Degrees of Belief." In *Prospects of Pragmatism*, ed. D. H. Mellor. Cambridge: Cambirdge University Press.

———. 1981. "Tractarian Nominalism." *Philosophical Studies* 40: 199–206.

Skyrms, Bryan, and W. L. Harper (eds.). 1988. *Causation, Chance, and Credence*. Dordrecht: Kluwer.

Slote, M. 1978. "Time in Counterfactuals." *Philosophical Review* 8: 3–27.

Smiley, Timothy. 1983–1984. "Hunter on Conditionals." *Proceedings of the Aristotelian Society* 84: 241–249.

Sobel, Robert. 1997. *For Want of a Nail: If Burgoyne Had Won at Saratoga*. London: Greenhill Books.

Sorensen, R. A. 1992. *Thought Experiments*. Oxford: Oxford University Press.

Sosa, Ernest. 1967. "Hypothetical Reasoning." *Journal of Philosophy* 64: 293–305.

——— (ed.). 1975. *Causation and Conditionals*. London: Oxford University Press.

Spade, Paul Vincent. 1982. "Three Theories of *Obligations*." *History and Philosophy of Logic* 3: 1–32.

———. 1992. "If *Obligationes* Were Counterfactuals." *Philosophical Topics* 20: 171–187.

Stalnaker, Robert C. 1968. "A Theory of Conditionals." In *Studies in Logical Theory*. American Philosophical Quarterly Monograph Series, no. 2, pp. 98–112. Oxford: Blackwell. Reprinted in Sosa 1975, pp. 165–179; in Harper, Pearce, and Stalnaker 1981, pp. 41–55; and in Jackson 1991, pp. 28–45.

————. 1970. "Probability and Conditionals." *Philosophy of Science* 37: 68–80.

————. 1975. "Indicative Conditionals." *Philosophia* 5: 269–286.

————. 1981. "A Defense of Conditionals Excluded Middle." In Harper, Pearce, and Stalnaker 1981, pp. 87–104.

————. 1984. *Inquiry*. Cambridge, Mass.: MIT Press.

Stalnaker, R. C., and R. H. Thomason. 1970. "A Semantical Analysis of Conditional Logic." *Theoria* 36: 23–42.

Stalnaker, R. C., and Richard Jeffrey. 1994. "Conditionals as Random Variables." In Eells and Skyrms 1994.

Stevenson. C. L. 1970. "If-iculties." *Philosophy of Science* 37.

Stewart, H. F., E. K. Rand, and S. J. Teolos (eds.). 1973. *Boethius' Theological Tractates, and Consolations of Philosophy*. Cambridge, Mass.: Harvard University Press.

Strawson, P. F. 1952. *Introduction to Logical Theory*. London: Methuen.

————. 1986. "'If' and '⊃'." In *Philosophical Grounds of Rationality*, ed. R. E. Grandy and R. Warner, pp. 229–242. Oxford: Clarendon Press.

Swain, M. (ed.). 1970. *Induction, Acceptance, and Rational Belief*. Dordrecht: Reidel.

Sylvan, Richard, and J. Norman (eds.). 1989. *Directions in Relevant Logic*. Dordretch: Reidel.

Teller, P. 1973. "Conditionalization and Observation." *Synthese* 26.

Thomason, Richmond. 1970. "A Fitch-Style Foundation of Conditional Logic." *Logique et Analyse* 52: 397–412.

Thomason, R. H., and Anil Gupta. 1981. "A Theory of Conditionals in the Context of Branching Time." In Harper, Pearce, and Stalnaker 1981.

Thompson, James F. 1954. "Tasks and Super-Tasks." *Analysis* 15: 1–13. Reprinted in R. M. Gale (ed.), *The Philosophy of Time* (London: Macmillan, 1968).

————. 1990. "In Defense of ⊃." *Journal of Philosophy* 87: 56–70.

Thomsen, Brian, and Martin H. Greenberg (eds.). 2002. *Alternate Gettysburghs*. New York: Berkeley Books.

Tichy, Pavel. 1976. "A Counterexample to the Lewis-Stalnaker Conditional Analysis of Counterfactuals." *Philosophical Studies* 29: 271–273.

————. 1978. "A New Theory of Subjunctive Conditionals." *Synthese* 37: 433–457.

Traugott, Elizabeth Closs, C. A. Ferguson, and J. S. Reilly (eds.). 1986. *On Conditionals*. Cambridge: Cambridge University Press.

Tsouras, Peter G. 1997. *Gettysburg: An Alternate History*. London: Greenhill Books.

————. 2000. *Disaster at D-Day: The Germans Defeat the Allies, June 1944*. London: Greenhill Books.

Unger, P. 1982. "Toward a Psychology of Common Sense." *American Philosophical Quarterly* 19: 117–129.

van Fraassen, Bas C. 1976. "Probabilities of Conditionals." In Harper and Hooker 1976, pp. 261–308.

————. 1981. "A Temporal Framework for Conditionals and Chance." In Harper, Pearce, and Stalnaker 1981, pp. 332–340.

————. 1982. "Epistemic Semantics Defended." *Journal of Philosophical Logic* 19: 463–464.

van Inwagen, Peter. 1989. "Indexicality and Actuality." *Philosophical Review* 89: 403–426.

Veltman, Frank. 1986. "Data Semantics and the Pragmatics of Indicative Conditionals." In Traugott et al. 1986.

Walker, R. C. S. 1975. "Conversational Implicatures." In *Meaning, Reference, and Necessity*, ed. S. Blackburn. Cambridge: Cambridge University Press.

Wakker, Gery. 1994. *Conditions and Conditionals: An Investigation of Ancient Greek.* Amsterdam: J. C. Gieben.

Walters, P. S. 1961. "The Problem of Counterfactual Conditionals." *Australasian Journal of Philosophy* 39: 30–46.

———. 1967. "Counterfactual Conditionals." In *The Encyclopedia of Philosophy*, vol. 2, ed. P. Edwards, vol. 2, pp. 212–216. New York: Macmillan.

Warmbrod, Kenneth. 1983. "Epistemic Conditionals." *Pacific Philosophical Quarterly* 64: 249–265.

Watling, J. 1957. "The Problem of Contrary-to-Fact Conditionals." *Analysis* 17: 73–80.

Weinberg, Julius. 1951. "Contrary-to-Fact Conditionals." *Journal of Philosophy* 48: 17–21.

Weiner, J. 1979. "Counterfactual Conundrum." *Noûs* 13: 499–509.

White, David. 1985. "Slippery Slope Arguments." *Metaphilosophy* 16: 206–213.

Will, Frederick L. 1947. "The Contrary-to-Fact Conditional." *Mind* 56: 236–249.

Williamson, C. 1969. "Analyzing Counterfactuals." *Dialogue* 8: 310–314.

Wilson, Fred. 1986. *Laws and Other Worlds.* Dordrecht: D. Reidel.

Wiredu, J. E. 1971. "Material Implication and 'if . . . then.'" *International Logic Review* 3.

Woods, Michael. 1997. *Conditionals.* Oxford: Clarendon Press.

Worley, Sara. 2002. "In Defense of Counterfactuals." *Philosophia* (Israel) 29: 311–323.

von Wright, G. H. 1957. *Logical Studies.* London: Routledge.

Young, J. J. 1972. "Ifs and Hooks: A Defence of the Orthodox View." *Analysis* 33.

Zalta, Edward N. 1988. *Intensional Logic and the Metaphysics of Intentionality.* Cambridge, Mass.: MIT Press.

Ziehen, Theodore. 1920. *Lehrbuch der Logik.* Bonn: A. Marcus & E. Weber.

Name Index